SPANNING THE GATE

The Golden Gate before the Bridge.
—Ansel Adams

Baron Wolman
presents

SPANNING THE GATE

The Golden Gate Bridge

Text by Stephen Cassady

Squarebooks
Mill Valley, California

Photo Credits. For the most part, the photographs appearing in this book are from negatives in the archives of the Golden Gate Bridge, Highway and Transportation District. Several others came from the collection of the Redwood Empire Association. The photograph on page 2 is by Ansel Adams, the one on page 101 is from the California Historical Society, the ones on page 126 are from the San Francisco Examiner and from International News Photos. Enlargements from large format negatives were made by Edward Dyba of San Francisco. Prints from 35mm negatives were made by Chong Lee of San Francisco. Hand-tinting of the black and white photographs was done by Ann Rhoney of San Francisco. The cover photograph was hand-tinted by Alba Vasconcelos. Jack Doonan of San Francisco retouched the prints where necessary.

The first edition of *Spanning the Gate* was designed by Clyde Winters. The re-design for this edition was done by Lauren Kahn of Cassidy & Company.

Published by Baron Wolman/SQUAREBOOKS

Library of Congress Catalog Card Number 77-83284 (First Edition)

ISBN: 916290-35-2 (Revised Edition)

10 9 8 7 6 5 4 3 2 1

Revised Edition

SQUAREBOOKS, Inc.
Post Office Box 1000
Mill Valley, CA 94942

**Spanning the Gate is dedicated to
those who built, maintain, cross and
cherish the Golden Gate Bridge.**

Preface: Above The Gate

The bridge went up in the Depression, 1933 to 1937, when six lanes for auto traffic seemed more excessive than adequate, when spanning the Golden Gate channel seemed as impossible as building the Pyramids once had to every Egyptian save Cheops.

Decades have elapsed since construction and opening day, decades in which the Golden Gate Bridge has become one of the most visited tourist attractions in the United States. Decades proving correct the vision of chief engineer Joseph Strauss, proving wrong the naysayers, those in the beginning who said the bridge would not endure—if indeed it could be built at all, and then at a cost that would not bankrupt the participating north coast counties.

The story of the Golden Gate Bridge comes in two parts. The first is a political saga beginning in 1910—when the idea for a span connecting San Francisco with Marin first gained public support—and ending in 1933 with a groundbreaking ceremony.

The hero of the first tale is Joseph Strauss himself. A bridge would have been built eventually. But without Strauss's indefatigable sense of mission, it would not have been erected in the monumental way it was, not with such a fine blend of adventure, economy, function, and art.

The Golden Gate Bridge that Strauss first envisioned would be the longest and tallest suspension span ever constructed (it was not a matter of choice; the hydrophysics of the Golden Gate channel dictated the dimensions). It was a daring leap into the unknown, an engineering feat without precedent, the only major public project ever built in the United States without state or federal aid.

Opponents were everywhere, and Strauss fought them all to a decision in his favor. He battled politicians, citizen-taxpayers, financial lobbyists, big-business pressure groups, rival engineers, and geologists. He won out, of course, but according to his son, the strain on his heart eventually killed him.

The second story, the construction itself, 1933 to 1937, is a tale of diverse heroes blending into a single blue-collar entity: the collective builder—workers and engineers who constructed the span in bone-chilling conditions. "The coldest damn job I ever worked on" was the consensus of bridgemen, many of whom had put up steel all over the world. Against odds natural and otherwise, they brought in the bridge on time without cost overruns or defects.

"The most beautiful bridge in the world," said a worker still proud in 1985, a man in his eighties named Ed Davenport. "Especially that color. That was Strauss. He invented that color. Up to then, every damn bridge we worked on was battleship gray."

Strauss pushed unyielding until the project was real. The workers and on-site engineers took over from there. With their courage, pride, and zeal, they elevated the battleship gray of routine bridgebuilding into the international orange of the Golden Gate.

We came to both stories in 1976, when the bridge was approaching its fortieth anniversary. San Francisco designer Clyde Winters opened the way. Digging in bridge archives, Winters had found some 10,000 negatives forming a photographic record of construction from groundbreaking to opening day. Most of the photos had never been published; the rest had not been seen in forty years.

The photos gave us threads that research would treadle through the loom. One thread took us back to Emperor Norton's whimsical suggestion for a bridge in 1872, another to a more serious campaign in 1916, a third to Strauss's involvement, which began in the spring of 1919 and lasted until opening day in late May of 1937.

Other threads led us to the workers, the men who braved storms and cold, who balanced themselves on six-inch I-beams while toiling hundreds of feet above the open sea, who on one fateful day in 1937 watched in horror as ten of their brethren fell into the "hole," as they called it. They looked back at a fulfilling time, men in their seventies with recall intact, most of them eager to pass along their stories. Ed Davenport was one of them.

Davenport was the chief inspector for the steel work. He had been imported for the job from Philadelphia by resident engineer Russ Cone. "I was painting a bridge back there," he says. "I was lucky even to have a job. My home was in Kansas City. The district here had a rule: you had to be a resident of one of the seven counties in order to work on the Golden Gate Bridge.

"But when it come time to put up steel, Cone went to the Board of Directors. He said, 'There's not a man in the whole seven counties that's even seen a suspension bridge, let alone worked on one.' He said, 'We got to have someone who knows suspension bridges to be an inspector—take charge of the steel inspection.' They asked if he had anyone in mind. He named me. I came out in '33."

Davenport also moonlighted as a diving inspector when the sub-contractor, Pacific Bridge, was pouring concrete for the south tower.

"Cone had a professional diver doing his inspection," Davenport says. "I'd had some diving experience. I said, 'Hell, I want to get in on this.' I knew Cone was coming over to the Marin side one day. I left a picture on my desk—of me in a diving suit, with a helmet on my lap. He saw that picture and got me on the phone. 'Why didn't you tell me you were a diver? Get the hell over here! I want an engineer doing this inspection, not a goddamn diver.'

"That was okay with me. They'd pay me 25 bucks a dive, in addition to my salary. Hey, I needed the money. This was the Depression. My son had been born on the east coast, and I hadn't paid for him yet. Those dives paid for my son."

Ed Davenport was typical—he was one of the workers—and *Spanning the Gate* is their story. Those who built a bridge that would last a half-century and beyond: from the anonymous operator of the steam shovel that first broke ground on the Marin shore in 1933 to the bridgeman who pounded in the golden rivet on opening day in 1937, a man named Iron Horse Stanley.

"That guy was something," marvels Ed Davenport, whose crew inspected every rivet on the bridge. "Best worker I ever saw. Riveting crews had quota of 350 in eight hours. Stanley would meet his quota, all perfectly done, by two in the afternoon, then go home.

"He was something." They *were* something, all of them.

Stephen Cassady
Merced, California
June, 1986

Prologue: The Open Sea

Building a bridge is a war with the forces of Nature.
—Joseph Strauss

The Wrath of Nature

San Francisco Bay extends some 52 longitudinal miles from the marshlands of the San Pablo shores in westernmost Solano County to Coyote Creek in the Dumbarton shallows at the top of the fruit-rich Santa Clara Valley. About two-fifths the distance from north to south, the Bay joins the open sea through a five-mile-long cleavage carved through California's coastal highlands.

As the only path between ocean and bay, the Golden Gate grants passage to both fresh waters and marine. Seabound rivers rushing in from the east can exit only through this narrow aperture in the coast. Heavy swells off the Pacific have this same mile-wide channel as their only point of entry. In forceful tides of ceaseless alternation, they pour through the Gate four times a day, powerful and compressed, hostile waters often frothed to rage.

Fresh waters reach the Golden Gate through meandering paths they have cut in the mountains and valleys of inland California. The Sacramento River is formed from high glacial streams scouring the slopes of Mount Shasta 400 miles northeast of the Gate. This infant river flows down into Lake Shasta, where it meets with waters from the McCloud and the Pit and emerges mature for its journey through the broad alluvial plains of the Sacramento Valley. The Sacramento reaches high water near the upper valley town of Red Bluff, then makes its way southerly through the rice paddies, wheat fields, and flowering fruit orchards of the boundless green and dun quiltwork of flatland agriculture. Along the way, it is joined by fast rivers riffling down from the Sierra Nevada: the Feather, the Yuba, the Bear, the American. Engorged with their waters, the big river makes a cuphandle turn just above the city of Sacramento and heads west to San Francisco Bay.

Several hundred miles to the south, the San Joaquin originates in two forks on the eastern slopes of the Sierras. It travels across the mountains and down through the cattle-dotted foothills of Madera County, southwest at first, until well into the fertile San Joaquin Valley. At the geographical center of the state, near the melon fields of Mendota, the San Joaquin changes course and heads northwest for 125 miles, paralleling the endless string of valley farm towns, absorbing rivers which flow down from the mountains, the Fresno, the Merced, the Stanislaus, the Tuolumne, the Mokelumne, the Cosumnes. Near the inland port city of Stockton, the San Joaquin bends west, eventually meeting the Sacramento at the delta hamlet of Collinsville. Together they form the 425,000-acre San Joaquin-Sacramento Delta basin.

From Collinsville, the confluent rivers rush unimpaired into San Francisco Bay, an average of 22.2 billion acre-feet annually surging past Pittsburg and Port Chicago, into Suisun Bay east of Vallejo, through the Carquinez Straits between Benicia and Martinez, and into the Bay of San Pablo on the north end of San Francisco harbor. These two great fresh water arteries, their principal veins and myriad anonymous capillaries drain over 59,000 square miles of area, close to 40 per cent of the acreage of the state of California. Joined by runoff from seventy-seven local streams with names like Alameda Creek, Napa River, Coyote Creek, Sonoma River, Guadalupe River, Petaluma River, and Walnut Creek, this massive transport of seabound fresh water adds heavily to the bloating of the Bay, which in the ebb tide of the Pacific rushes through the narrow throat of the Golden Gate with such uncontrollable force that it empties one-sixth its contents every twelve hours at an average flow seven times that of the Mississippi River.

Ocean waves sweeping into the Golden Gate begin in ripples as far away as Antarctica and the South Pacific, and are driven to swell by eastbound winds which carry them on their long journey across the sea. Waves are capable of traveling great distances, up to thousands of miles, and are stopped only by the cessation of wind or the intrusion of land.

Approaching a coast, waves will rise up and break when their height is roughly eight-tenths the depth to the sea bottom beneath them—in other words, when they feel from the bottom that a beach is near. Against an unbroken coastline and smoothly sculpted sea floor, ocean waves will hit land in orderly columns which crest uniformly then recede as troughs of expended foam.

The Pacific waves assaulting the mouth of the Golden Gate have benefit of neither continuous coastline nor level sea floor. The channel varies radically in depth, from 43 to 313 feet, increasing rather than absorbing the mounting wave energy. Where nearby beaches dissipate the traveling sea with their gently sloping shores, the Golden Gate pulls it toward the Bay with jetstream force. Nor do the waves rush the channel in any uniform manner. Baybound currents churn in constant confusion from premature shallows lying just outside the Gate. Extending from the beach just below the San Francisco Zoo, a crescent-shaped sand bar swings in a wide arc all the way across the mouth of the gorge to Rodeo Cove on the Marin Peninsula. The bar varies in depth from 23 to 35 feet, except for a shipping lane in the middle of the arc which has been dredged half again as deep as the bar itself.

Some waves flow unbroken through the shipping channel, skating through the gorge in swift but untroubled currents, while myriad others are feeling bottom at differing points and angles along the bar. With every flood tide, countless waves of varying heights are swelling, breaking, shoaling, and refracting just before they reach the channel. They flow into the Gate as a turmoil of riptides, whirlpools, eddies, and boils, raging uncontrolled until they pass through the narrows between Lime Point in Marin and Fort Point in San Francisco and are tranquilized by mixing with the calmer waters of the Bay.

All this flowing and ebbing force is concentrated at the narrows of the Gate, where the channel is deepest and the width of the gorge is most at odds with the mass of water pouring through. Here the channel never has a chance to relax. Twice a day, gravitational attraction from the moon raises the powerful Pacific currents and floods them redoubled between Lime Point and Fort Point; twice a day, it pulls them into retreat for compressed exit from the river-bloated Bay. Here periods of slack water may last only 20 minutes between alternating tides that hurl 2,300,000 cubic feet of water per second through a one-mile wide channel at speeds between 4½ and 7½ knots an hour. Here the Gate is a wind tunnel framing 20- to 60-mile-per-hour gusts which change directions with the seasons but remain constant and corrosive January through December.

Here, too, despite the wrath of Nature which made the task seem impossible, is the only imaginable location for a bridge between San Francisco and Marin.■

PREHISTORY: THE DREAM PERSISTS

It took two decades and two hundred million words to convince people the bridge was feasible.
—Joseph Strauss

The Civic Challenge

In its infancy San Francisco was a sparse settlement under Spanish rule, a small community as remote as Tortuga located along the sand dunes on the northeast rim of the San Mateo Peninsula. It was founded as the village of Yerba Buena in the spring of 1776. At first, Yerba Buena consisted of a presidio under no threat of invasion, and a mission for evangelizing native Indians too docile to resist. For the first seven decades of its existence, Yerba Buena languished under the neglect of its parent governments in Madrid and Mexico City. In the 1840's things began to change.

Progress arrived with the merchant ships that berthed in the Bay to trade for hides and tallows produced on nearby rancheros. Shippers found they could avoid the heavy tariffs levied in the California capital of Monterey by conducting business in Yerba Buena instead. Soon towns sprang up along the shores like wild grass following a spring rain.

The region prospered even more after California was taken from Spain in 1848. When the treaty of Guadalupe-Hidalgo was signed in February of that year, San Francisco Bay was grafted onto an already frontier-conscious America. All that was needed then was a thunderclap to stampede people west. It occurred just eight days before Guadalupe-Hidalgo, when gold was discovered at Sutter's Mill in the Sierra foothills above Sacramento. Prospectors, settlers, profiteers, and merchants rushed into Northern California, ambitious for wealth and gain. By the end of 1849, Yerba Buena, once a village of fewer than 400 inhabitants, became San Francisco, a city of 35,000.

The story of the Golden Gate Bridge reaches back to the mid-19th century, when it became apparent that San Francisco, surrounded on three sides by water, would soon need closer connection with the satellite regions to the north and the east. Growth all along the Bay was without foreseeable limit, so the matter of a bridge was only a question of time—the time it would require technology and vision to overtake necessity. Strangely, recognition came first from a madman.

Joshua Norton was a failed Gold Rush merchant who reacted to bankruptcy by losing his mind. Although destitute, he managed to outfit himself in imperial garb and became self-appointed

"Norton I, Emperor of the United States and Protector of Mexico." Norton rose quickly to status as a regional character. He financed his illusory empire with currency of his own printing which, along with his eccentricities, was accepted graciously in San Francisco.

Norton built a reputation for lunatic pronouncement, so it came as no surprise that, in August of 1869, he had this edict printed in the Oakland *Daily News*: "We, Norton I . . . do order and direct that a suspension bridge be constructed from . . . Oakland . . . to Yerba Buena Island, from thence to the mountain range of Sausalito, and from thence to the Farrallones, to be of sufficient strength and size for a railroad." With that, the idea for a bridge across San Francisco Bay was a matter of public record.

The first mention of crossing the Golden Gate came three years later. Railroad potentate Charles Crocker foresaw San Francisco as the western terminus for his Central Pacific line. Crocker, who had personally supervised the construction of the transcontinental railroad from Sacramento, California to Promontory Point, Utah had company engineers prepare detail maps and estimates for a suspension span from Marin to San Francisco. These he presented to the Marin County Board of Supervisors in 1872. But his plan was abandoned when Central Pacific engineers discovered instead a way to transport railroad cars across the Carquinez Straits by steam ferry.

The notion of a Golden Gate bridge lay dormant for more than four decades. But in 1916 it was revived by a man who had attended that 1872 board meeting. James Wilkins had been raised from the age of seven in the Marin County town of San Rafael. After graduating from the University of California at Berkeley with a degree in engineering, he abandoned his college major for newspaper work. He first ran a small periodical in Marin (*The Tocsin*). In 1910 he took a job with the now defunct *San Francisco Bulletin*.

Like all Marin County residents who worked in the city, Wilkins commuted daily across the Bay. Ferry travel in the second decade of the 20th century was equivalent to the horse and buggy, sufficient before the outbreak of motor cars, irritatingly slow for an age of mechanized land travel. By 1916, Wilkins was among a wave of people growing impatient with ferries. He was also among the few who realized that Marin's development would depend entirely upon closer association with San Francisco.

In August of 1916 Wilkins began an editorial campaign for a bridge across the Gate. "The northern counties," he wrote in his initial article in *The Bulletin*, "almost an empire in extent, with potentialities barely surface-scratched, contain a present population that has passed well beyond the two thousand mark. Nature has, in a way, tied their fortunes, beyond recall, to San Francisco. That city must always be their final marketplace—the clearing house. Nothing can be more important to San Francisco, beset as it is by jealous rivals carelessly endeavoring to win away its trade, than the speedy development of a region whose business comes to it automatically, which can never be diverted elsewhere. And nothing can hasten that development more than the free circulation of modern life by a bridge across the Golden Gate."

Wilkins also appealed to man's timeless urge for spectacular challenge. "Even in the remotest times, long preceding the Christian era," he wrote, "the ancients understood the value of dignifying their harbors with impressive works. The Colossus of Rhodes and the Pharaohs of Alexandria were counted among the seven wonders of the world. The same tendency appears in our own times, witness the cyclopean statue at the entrance of New York harbor. But the bridge across the Golden Gate would dwarf and overshadow them all."

Drawing on his engineering background, Wilkins then wrote about the bridge he envisioned. It would be a suspension span about 150 feet above high water, high enough to clear the tallest ship mast. It would be located between Lime Point in Marin and Fort Point in San Francisco. It would have a center span of 3,000 feet and two side spans of 1,000 feet each. The entire structure, he estimated, would cost approximately $10,000,000. Wilkins' technical grasp did not extend far beyond this rudimentary plan, but in concept he was astoundingly close to the bridge that was still twenty-one years away from being built.

The *Bulletin*'s articles began to stir interest, notably within the mind of San Francisco city engineer M. M. O'Shaughnessy. Just two years earlier, O'Shaughnessy had mobilized the $7,000,000 Hetch-Hetchy project, a system of dams and aqueducts for transporting water 156 miles from Yosemite Valley to the chronically dry city of San Francisco. O'Shaughnessy would be the first civic authority to promote the spanning of the Gate, the first practicing engineer to sense that it was possible. He would withdraw his support some years later, after questioning the wisdom of bridging the Gate during the Depression, but his early enthusiasm gave bridge backers a first firm bond between dream and reality.

From its very beginning the Golden Gate bridge was slow to develop. O'Shaughnessy was unsure about the project's material expense and its structural feasibility. And in the fall of 1916 he was diverted from any comprehensive investigation by the larger issues facing all of America, the war in Europe and the election of a new president. It was not the time for parochial concerns. By the end of the calendar year 1916, the Golden Gate crossing was just a small flame flickering in the minds of two men.

Still, O'Shaughnessy could not completely set aside his fascination. When possible during the war years, he polled engineers of his acquaintance for their opinions of a Golden Gate bridge. Most of the replies were negative, particularly in terms of cost, with some of the rough estimates running as high as $100,000,000.

Only one reputable engineer reacted positively. A grim, Teutonic Chicagoan named Joseph Strauss was in San Francisco often in 1917, consulting with the city engineer on the construction of some waterfront street bridges. As an aside, O'Shaughnessy asked him to evaluate Wilkins' ideas. "He accompanied the inquiry with the statement that 'Everybody says it can't be done, and if it could be done, it would cost over $100,000,000,' " Strauss wrote in 1937, in a first-person account for the *Sausalito News*. Undaunted, Strauss said offhandedly that the Gate could be bridged, and for a reasonable sum, though it would take some study to determine the particulars of either claim.

Nothing more was done until after the Armistice, but late in November of 1918 civic appetite for challenge was whetted once more by Wilkins' tantalizing columns. To meet what it considered growing sentiment, the San Francisco Board of Supervisors, on a resolution introduced by member Richard J. Welch, called upon Congress to authorize a federal survey of the Golden Gate channel. Nine months later, in August of 1919, the Board sent a letter to city engineer O'Shaughnessy officially directing him to explore "the subject of the proposition of building a bridge across the Golden Gate . . . no expense to be incurred hereunder."

As one of his first official gestures O'Shaughnessy asked Strauss and, unbeknownst to Strauss, two other engineers for calculated estimates of a Golden Gate crossing. Strauss received the challenge with enthusiasm. In return he requested only a budget ceiling and a hydrographic study of the channel. O'Shaughnessy talked of $25,000,000 to $30,000,000; and he reported the Board's resolution of November 1918. Assured on those points, Strauss was ready to begin.

He was not to receive the technical data from the U.S. Coast and Geodetic Survey until the spring of 1920, but in 1919 he personally reconnoitered the area. From Crissy Field at the Presidio he hiked up the hill to the promontory overlooking Fort Winfield Scott, the pre-Civil War edifice standing sentinel over the narrows of the Gate. There he stood, like a bowsprit on the very tip of the San Mateo Peninsula, an implacable little man looking out at the blue-green, foam-flecked eddies and rips that rush unrestrained through the channel. He was seeing firsthand the challenge he would face building a bridge across the open sea.

Joseph Baermann Strauss was an odd amalgam of qualities. He was 49 years old in 1919, a man of real talent, given to peevish fits of temper. "You didn't work for him for the compliments you might receive," recalls Ruth Natusch, public stenographer for the Palace Hotel, who did much of Strauss's secretarial work in the early years of the bridge project. "In fact, the girls in my office would hide when they saw him coming." Once, when working on a monorail design in Los Angeles, Strauss so exasperated the public secretary there that she threw her steno pad at him and stalked out the door.

"He was gruff and forbidding," says Miss Natusch. "He would mumble when he dictated. He would talk in complex engineering terms without bothering to explain them. And he was very short-tempered. But for some reason—I guess it was that I was brought up to be obedient—I loved working for Mr. Strauss. It was mesmerizing just to be around somebody who was accomplishing what he was."

Like most men of historical stature, Strauss had a dreamer's love of destiny. His imagination soared sometimes without rein, probably the legacy from a musician mother and an artist father. Miss Natusch remembers that on many occasions when she was asked to transcribe her employer's notes, she would find poems jotted absently among the technical theorems and equations.

San Francisco writer Harold Gilliam said of Strauss, "As a boy in Cincinnati, [he showed] an early talent for writing romantic poetry. But he also had a bent for mathematics and mechanics. The two sides of his character, the imaginative and the practical, became the source of his genius. When the engineer in him warned that something could not be done, the poet challenged him to do it anyway. And usually the engineer found a way."

Joseph Strauss was a man of overwhelming ego. He argued hotly and often with his assistant Clifford Paine, sometimes about matters as minute as the wording on the masthead plaque for the finished bridge. Paine wanted his name bronzed as "Principal Assistant Engineer." Strauss would have had it read "Principal Assistant to the Chief Engineer."

When he saw the opening of the rival San Francisco-Oakland Bay Bridge spread out in headlines and stories a full six months before his own bridge was finished, he was said to have snorted, "Bridge, hah! That's no bridge, that's a trestle."

It was as though Strauss were beset by demons, all striving for dominance while running wildly between the petty and the sublime. Fortunately for the successful denouement of the two-decade drama of joining San Francisco with Marin, they were all willing to ally for the sake of one great immortal challenge.

Indeed, the man to bridge the Golden Gate could not have been an ordinary civil engineer. The task demanded a visionary smitten by the idea of linking the western coastline at its only overland fissure; a technician challenged by problems that had confounded engineers since 1872; an evangelist capable of convincing an indifferent public of the need for a bridge; a politician who could mobilize legislators, financiers, the military, and the electorate all into a unified cause; and, last, a pillar of psychological strength who could remain motivated through fifteen years of obstacles which would have driven a lesser human to lunacy, or at the very least to failure.

Strauss had been a bridge romantic since growing up in the shadow of a great span. The Strauss family home looked out upon the high stone towers of the "Biggest Bridge in the World," the Cincinnati-Covington, which crossed the Ohio River into Kentucky. The Cincinnati Bridge was opened in December of 1866, three years and a month before Strauss was born. With its 1,057-foot main span suspended gracefully from great steel cables, it became a backyard totem for Strauss's mechanical inclinations. According to legend, it presented him also with the first sharp vision of his future.

As an undersized sophomore at the University of Cincinnati (he was just over five feet tall as an adult, weighing 120 pounds), Strauss tried out for the university football team. The hulking upperclassmen showed him little mercy; predictably, he was injured almost as soon as he took the practice field for the first time. The injuries were serious enough to require hospitalization, and it was in the college infirmary that Strauss allegedly underwent his conversion. He was seduced by engineering while gazing out the hospital window at the Cincinnati Bridge. If football had belittled him, the legend reports him to have thought, bridge building would raise him up again.

The story may have been embellished by several decades of retelling; nonetheless, Strauss did change his major to engineering, and it proved the seamless weld of a man with his destiny. The engineering curriculum exposed a hidden genius for mathematics. At the same time, it framed his imagination within the rigid disciplines of science.

Strauss took his degree in engineering in 1892. For his graduation thesis he presented a design for a bridge joining the continents of North America and Asia at the 50-mile-wide Bering Straits. It was a fanciful creation, more the work of the dreamer than the technician, and it survives only as an artifact in a university library. But his ambitious plan for bridging two continents was the first evidence that Joseph Strauss possessed a unique mind. "Our world of today," he would say more than a quarter of a century later, trying to sell the Bay Area populace on the idea of a Golden Gate bridge, "revolves around things which at one time couldn't be done because they were supposedly beyond the limits of human endeavor. . . . Don't be afraid to dream."

Strauss did not distinguish himself in his field until after the turn of the century. In 1902, while working as office manager for the Ralph Modjeski firm of Chicago, he submitted an unsolicited alternative to some current designs.

Modjeski was among those firms growing dissatisfied with the turnstile types of moveable bridge, thinking them too space-consuming for modern navigation. The designers were beginning to pay greater attention to the bascule bridge, more or less the descendent of drawbridges that were raised and lowered over the moats of medieval castles. (Basically the bascule bridge operates on a seesaw principle: a single or double-leaf span hinged to the roadway is hoisted and dropped by mechanized counterweights.)

The Modjeski people embraced the bascule theory, but they were snagged on a functional difficulty. Balancing the leaf spans required massive pig iron counterweights as costly to produce as the spans themselves. In Strauss's design, the counterweights would be built of inexpensive concrete. He dealt with the fact that concrete is lighter than iron by redesigning the bascule mechanism to accommodate much more bulk. Strauss felt he had invented the ideal solution for a vexing problem. But when he showed his blueprint to the resident engineers, they dismissed it with a scoff.

Incensed that his idea could be so summarily rejected, Strauss resigned from Ralph Modjeski. A cautious man would not have made such a move, but Strauss was not intimidated by his sudden unemployment or the scant prospects for a new job. Reasoning that his original bascule design was marketable, he rented a small office in Chicago and opened his own firm.

At first, his independence seemed unwise; contracts came in slowly, if at all. Finally, he had to mortgage his very future just to establish the business. He contacted a railroad company in Cleveland which had been advertising for a moveable bridge across the Cuyahoga River near Lake Erie. The company was intrigued enough with Strauss's ideas to authorize a bascule rail bridge, but not enough to pay for it until completed. So Strauss assumed full risk for the project. He borrowed to the edge of bankruptcy and contracted to build his first significant bridge.

It cost $80,000 to cross the Cuyahoga River, but Strauss's bascule span proved so functional that it revolutionized the construction of moveable bridges. More, it netted sufficient capital and reputation to send the Strauss Company securely into the future. From the completion of that initial project to the beginning of his work on the Golden Gate, Strauss would build more than 400 bridges in the United States, Canada, and abroad. These included a railway span in Illinois whose support piers would be sunk in quicksand, 40 bridges in the mountains of Panama whose parts would be packed in by mule, and a double-leaf bascule bridge across the river Neva at the Winter Palace of the Tsar of Russia that would ultimately be stormed by rebellious peasants in the revolution of 1917.

By 1915, Strauss had expanded his operation both geographically and creatively. His mind would often travel away from bridges, following inventive notions, some destined to fail, but all searching and alert. A device to replace automobile tires with spring wheels failed, as did an airtram that would have predated the monorail. But they were overshadowed by the successes: a glass-washing machine for soda fountains, a bascule door for aircraft hangars, a concrete freight car for military transport, a portable searchlight that would be used widely by the Allies in World War I. And for the 1915 Panama-Pacific Exposition, held in San Francisco's Marina District just a few miles from the seething waters of the Golden Gate, Strauss's Aeroscope—a glass-enclosed platform attached to a crane which lifted observers 150 feet into the air—was a featured exhibit gaining him world-wide attention.

While he was building bridges and securing patents, Strauss based his firm in Chicago. Like many international engineering concerns which serviced the western United States and the Pacific, he also maintained offices in San Francisco's Palace Hotel.

With his vision, his background, and now his proximity, it was almost inevitable that Joseph Strauss would bridge the Gate. He was probably the one man in 1918 America Promethean enough to think the project within his grasp.■

Joseph Strauss was the Chief Engineer for the Golden Gate Bridge, but his contributions far transcend the title. Strauss was the first to question that the Gate was unbridgeable, and the first to submit a working blueprint. He led the fight against skeptical citizens and venal lobbyists. He helped to organize the political, promotional, and financial phases of the project. He oversaw its final design and construction. The finished bridge was a life's dream fulfilled.

ORGANIZATION: THE MEETINGS BEGIN

It is possible to bridge the bay at various points, but at only one can such an enterprise be of universal advantage—at the water gap, the Golden Gate—giving a continuous, dry-shod passage around the entire circuit of our inland sea. —James Wilkins

Manifest Destiny

From the time the city of San Francisco first reached metropolitan dimensions, people crossed the Bay exclusively by ferry. It was a distinct and charming method for travel, but one with a limited future. The ferry system was designed for the pedestrian passenger and, in lesser proportion, the horsedrawn carriage. Population increases in the Bay region, from 50,000 people in 1849 to 1,330,000 by World War I, imposed a pressing strain on the transbay system. The age of the automobile overloaded it to the point of collapse.

California citizens owned 44,000 autos in 1910. Nearly a decade later that figure had risen to 600,000, and there was no stopping the trend. 123,000 autos had ferried across the Golden Gate in 1919; in 1928, the number was 2,195,520. On Labor Day weekend, 1930, the Sausalito to San Francisco ferry line experienced the worst traffic snarl of its history. Motorists had to queue up on Sunday, and on Monday the situation boiled over. 9,300 vehicles swarmed the ferry facilities that day, all, it seems, at the same time.

By 7:00 PM cars were backed up three abreast for seven miles. Highway 101 was relatively clear for the next five miles, but at San Rafael, lines extended back into a 10-mile long bottleneck. By putting fourteen boats on the route and running them four minutes apart, the ferry officials managed to send the last of the weary motorists across by 12:30 AM on Tuesday.

Public support for a bridge was registered as early as 1912, when the San Francisco Board of Supervisors passed a resolution recommending a span between San Francisco and Oakland. The War Department in Washington killed that idea three years later, fearing that a Bay bridge would obstruct commercial navigation or impede the naval fleet in Alameda. As a result, auto crossings began outside the shipping lanes, following the model of the Dumbarton rail bridge erected on the southern rim of the harbor in 1910.

In 1927, the Dumbarton drawbridge was opened. That same year, a toll bridge spanning the Carquinez Straits between Crockett and Vallejo was also completed. In 1929, the San Mateo Bridge linking the mid-Peninsula with the East Bay city of San Leandro received traffic for the first time.

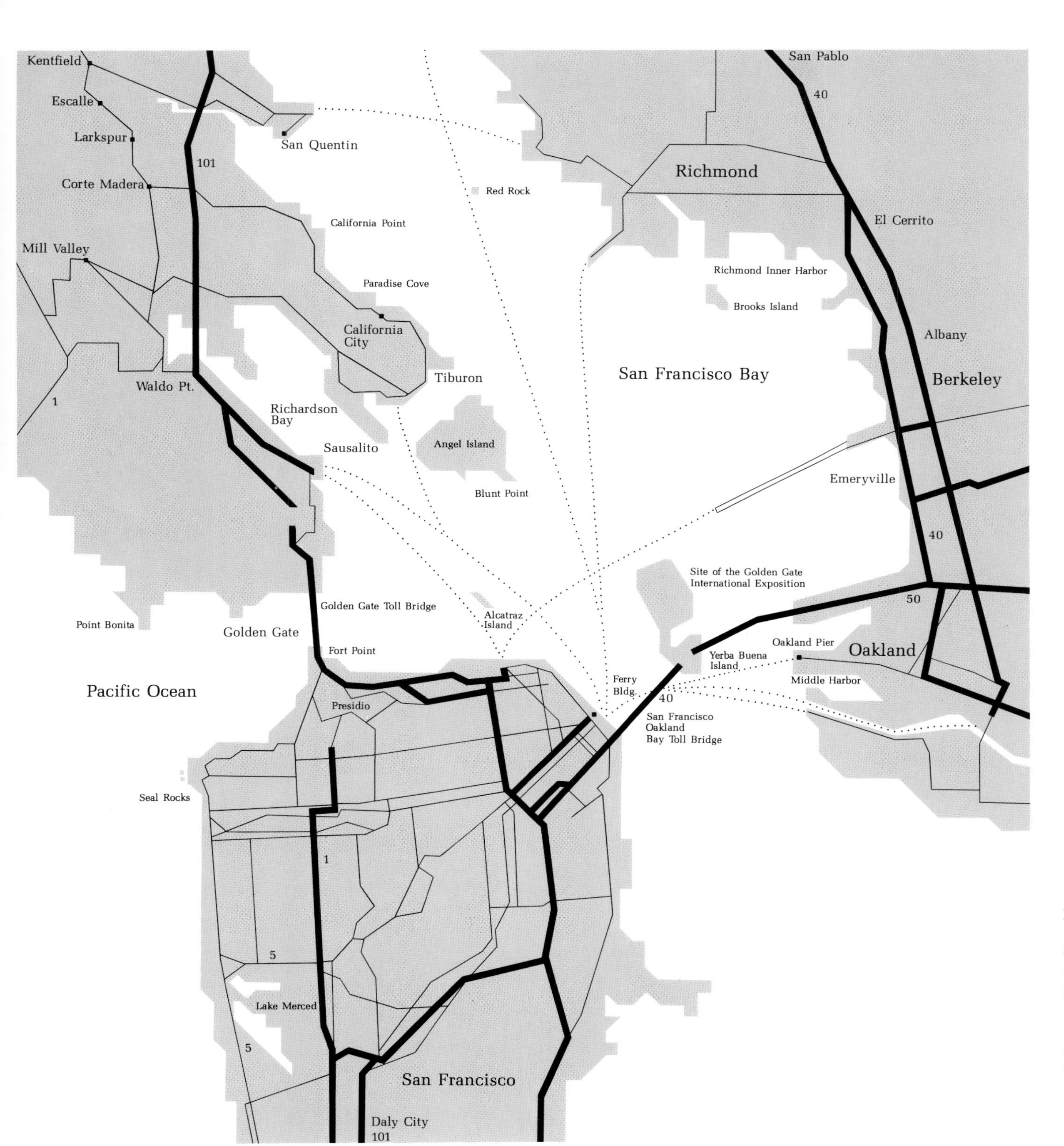

In the decade after World War I, Bay Area auto traffic increased seven-fold and strained the ferry system beyond its ability to cope. The San Francisco-to-Marin ferries were transporting as many as 3,000 autos during a single holiday weekend. But even with all boats in the water, any one-way route was capable of handling only 1,000 cars an hour, roughly the capacity of a dirt road during the horse-and-buggy era.

In 1929 also, a commission was appointed by former Bay Area resident Herbert Hoover to study and plan a San Francisco-Oakland bridge. Under its presidential seal, the Hoover-Young Commission developed a blueprint which satisfied all interests, and the Bay Bridge was begun in 1932.

A bridge across the Golden Gate, of course, fit into this evolutionary scheme, and it was with a local version of manifest destiny that Joseph Strauss and the bridge backers tried to promote the idea. But different from the others, this bridge proved a hard sell. The others were hailed as economically sound; this one as costly and misrepresented. The others would join San Francisco with highly developed areas; this one with sparse northern counties. It was pointed out—accurately enough, using ferry revenues as a guide—that current traffic and population did not warrant an expenditure of so many millions of dollars, particularly when there were so few guarantees that a Golden Gate bridge was feasible.

Widespread as it might have been, negative opinion was blind to 20th century progress. As the pro-bridge faction argued exhaustively, the horse and buggy era was forever dead; modern transport vehicles would sooner or later demand more and better arteries of travel; Marin needed only direct access to expand as dramatically as had any Bay community since the Gold Rush; and the opening of the Panama Canal in 1915 would readjust the world's trade routes, making the Bay region an important international crossroads worthy of the most sophisticated systems for travel.

In retrospect, it is clear that Strauss and his people were accurately assaying the future, but in the early 1920's their ideas were speculative and not a little frightening. They called for an expensive bridge underwritten by a district's power to tax property owners, a structure of unprecedented length built over waters thought unbridgeable, with nebulous costs based on the estimate of one ego-struck man from another community. There was such a swell of resistance, in fact, that fully fourteen years elapsed from that first supervisorial inquiry in 1919 to the breaking of ground at Fort Point in February of 1933.

When city engineer O'Shaughnessy was instructed by the San Francisco Board of Supervisors to proceed with bridge studies in late August of 1919, he stumbled immediately in a way typical of the prehistory of the bridge—he could not find an agency to survey the channel floor.

Finally, in January of 1920, he appealed to the U.S. Coast and Geodetic headquarters in Washington, D.C. USCGS Superintendent Lester Jones was sympathetic, and he authorized the USS *Natoma*, then stationed off the Pacific Coast, to sound the channel. O'Shaughnessy's office received the data recorded by the *Natoma* in May of 1920, whereupon the city engineer sent copies to three engineers, Joseph Strauss in Chicago; Francis C. McMath, President of the Canadian Bridge and Iron Company of Detroit; and Gustav Lindenthal, the man who engineered the 1,000-foot Hell Gate Arch over New York's East River in 1916.

McMath never officially replied, and Lindenthal disqualified himself by estimating that a Golden Gate bridge would cost a minimum of $56,000,000. Strauss, thinking he was the only one contacted, set out from the moment he received the survey to design a functional span within the $25,000,000 to $30,000,000 cost limit. It took him until June of 1921 to complete a design, a month and a half longer to compute its costs. But by August 3, O'Shaughnessy had in his possession a blueprint for a $17,000,000 Golden Gate bridge.

For some reason O'Shaughnessy failed to make Strauss's calculations public until December 7, 1922, almost a year and a half after he had received them. Part of the delay can be explained by revived interest in a San Francisco-Oakland crossing which occupied the city engineer and the Board until the plan was scrapped in January of 1922. And part was doubtless strategy developed by Strauss, O'Shaughnessy, and the other two members of the initial quartet of San Francisco backers, Supervisor Welch, and Mayor Rolph's secretary, Edward Rainey. These men admitted they thought it wise to build support on the north side of the Bay before approaching a reluctant public. It is possible, too, that O'Shaughnessy felt some trepidation over Strauss's design. If so, there were reasons.

Strauss had departed radically from the standard bridge forms in contemplation of the Golden Gate. He shared a reigning skepticism of the suspension bridge, particularly for this project, thinking it too supple a construction type to

In 1919, Joseph Strauss believed that a cantilever bridge would be too costly and heavy for the Golden Gate. And he felt that current metallurgy prohibited a suspension span longer than one-half mile. So he combined elements of both and invented a "symmetrical cantilever-suspension" hybrid. Considered grotesque when it was first made public in December of 1922, this rendering was nonetheless the design employed by early backers in their campaign for a Golden Gate crossing.

withstand the unabated fury which roared daily through the channel. But all other types of bridge, the arch, the truss, the cantilever, would require foundation piers that the depth of the channel simply forbade. Whatever his ultimate design, Strauss had to span approximately 4,000 feet between the shallows on either side of the gorge, more than twice the existing record of 1,800 feet held by the Quebec Bridge across the St. Lawrence Seaway. Meeting this kind of challenge called for innovative engineering, either a reassessment of old ideas, or the creation of new ones.

Characteristically, Strauss chose the latter course, inventing what he called a "symmetrical cantilever suspension" bridge, which had heavily girdered side spans extending 1,320 feet from both shores, leading to 800-foot-high towers. He shortened the untenable length of the center span by having cantilever arms reaching 685 feet from the towers toward mid-channel. Hanging from these cantilever arms, unsupported from beneath, would be a suspension center span of some 2,640 feet. "The cantilever by itself would have too large a self-weight for a span across the Gate," explained Strauss, "and so long a suspension bridge would lack the necessary rigidity and involve huge costs. The combination of the types uses the best features of each and admits an estimated cost of only seventeen and a quarter million dollars."

Strauss's ideas had backing even before he sent the drawings to San Francisco. As a preliminary test, he had shown them to three leading engineers in Illinois and New York. All responded favorably. "The plan seems to possess the virtues of both the suspension and cantilever types, and the vices of neither," said Charles Ellis of the University of Illinois Department of Engineering.

But despite such lofty assurances, nobody could say with certainty that Strauss's invention was sound. The problems in bridging the Gate were unique, and his solution had never been practically applied. In fact, it would prove technologically short-sighted. Within the next decade, advances in structural steel and changes in the theory of deflecting gale force winds would combine to make the suspension bridge the only reasonable alternative.

Strauss's original design had another drawback, and this one was immediately obvious. It was monstrous to look at, suitable perhaps for crossing some slag-ridden river in the industrial east, but offensive to the justly proud eyes of Bay Area citizens. Indeed, it was a graceless, overly girdered bridge, its cantilevered sides growing grotesquely out of the shores like horizontal oil derricks. If constructed, it would have stained one of Nature's sublime seascapes.

Still, it was the first tangible evidence that a bridge movement was afoot. And while the plans lay idle on O'Shaughnessy's desk, Strauss, with Rainey as a liaison, campaigned on the other side of the Bay. He spoke to business and civic groups, to citizens and legislators, anywhere he could stir public opinion. He was a poor speaker, monotonic and humorless, but his mind was fertile with vision, and that was enough to implant bridge ambition in people of influence. By the time his drawings were published in the local papers in December of 1922, Strauss had catalogued a long list of backers on the north side of the Golden Gate. In fact, all the initiative for organizing a bridge district derived from an area whose combined population numbered only 90,000.

By mid-January of 1923 the scattered support from both sides of the Gate began to ally. A banker from Santa Rosa, Franklin Pierce Doyle, invited all interested parties to attend a public meeting in the Santa Rosa city council chambers on January 14. Some 125 people showed up that late winter night, important people from San Francisco, the north counties, and the state legislature. Strauss was detained in Chicago, but San Francisco's Mayor Rolph, city engineer O'Shaughnessy, three supervisors and the president of the Chamber of Commerce were all there. The meeting opened with testimonials. O'Shaughnessy outlined the history of the movement, beginning with James Wilkins' articles in 1916. He told of the number of leading engineers who were enthused over Strauss's bridge design. Harry Speas, president of the Golden Gate ferry system, said he believed there would be business enough for both a bridge and the ferries, an opinion that would have ironic repercussions less than a decade later. A delegation from the State Assembly assured the group that congressional help would be forthcoming upon request.

The reports were long and encouraging, but incidental to the organization of the Bridging the Gate Association. The meeting began as an informal gathering of volunteer citizens; it ended with a cohesive group selecting an executive committee to pursue its aims. That historic first unit consisted of chairman W. J. Hotchkiss of Healdsburg, Captain I. N. Hibbard of San Francisco, Franklin Doyle of Santa Rosa, Frank L. Coombs of Napa, Richard Welch of San Francisco, Thomas Allen Box of Sausalito, and James G. Stafford of Santa Rosa. In addition, the Association appointed George H. Harlan, an attorney from Sausalito, unsalaried legal counsel and Joseph Strauss unsalaried engineering consultant.

Initially, the executive committee was concerned with public sanctions. Its first task was to approach the State Legislature for permission to form a legal district. The bridge movement needed the powers of taxation to raise money necessary for preliminary work on the project—channel borings, engineering reports, and the like; and also, the federal government granted construction permits over coastal waters only to public bodies established by state legislatures.

Frank Coombs was an Assemblyman from Napa, George Harlan a specialist in the organization of tax districts for such purposes as water and sewage systems. Together they set out to draft a bill for presentation to the Assembly in Sacramento. In the meantime, Strauss, Box and Hotchkiss went stumping through the north counties, urging boards of supervisors and city and county officials to pressure their congressional representatives into passing the Coombs Bill. The two missions coalesced in late May of 1923, when, after two months of Assembly debate, Governor Friend W. Richardson signed the Coombs Bill into law as the "Bridge and Highway District Act." The Association could now form a district to assume powers of taxation, eminent domain, building and maintaining roadways and bridges. With quick passage of the Coombs Bill, Strauss allowed himself a burst of premature optimism.

"The Golden Gate will be bridged and open for traffic by 1927," he vowed, "if the people of San Francisco and other communities of the Bay region are willing to spend $20,000,000 and if they are successful in obtaining the sanction of the War Department."

Acquiring approval from the Department of War threatened to be more than a formality. The federal agency had jurisdiction over all harbor construction that might affect shipping traffic or military logistics, and it owned the land on either side of the narrows of the Gate, the Presidio of San Francisco, and Fort Baker in Marin. In 1921 the Secretary of War had made his department's position plain. He killed a proposal for a San Francisco to Oakland crossing, in effect banning any construction north of the naval shipyard and drydock at Hunter's Point.

After the passage of the Coombs Bill there were hints that the military might relax its policy with respect to a Golden Gate bridge. But the Association could scarcely afford to rest its case upon gossip. For the better part of a year, it worked on an unimpeachable defense. Strauss subjected his design to exacting specifications; others from the Association conducted studies of traffic patterns. When the data was complete, they notified the San Francisco Board of Supervisors, which in turn forwarded the material to Washington. The War Department set May 16, 1924 as the date for an open hearing. It would be held in the chambers of San Francisco's City Hall and would be presided over by Colonel Herbert Deakyne, District Engineer for the U.S. Army.

A large crowd overflowed the chambers the night of May 16, all the backers from the 1923 organizational meeting, plus those since recruited. Anxiety was thick in the air, relieved only slightly by Deakyne's opening remarks. "There are only two points for the department to make its decision upon," he said. "The first and most important is whether or not the bridge will constitute a menace and a hindrance to navigation both in peace and war. The second is whether or not it will be adequately financed."

By now practiced in defense of the bridge, Strauss spoke for the Bridging the Gate Association. "San Francisco has often done the impossible," he intoned. "Now it only remains for her to connect up with the contiguous territory to make her the great city she is destined to be. Ways of transportation are essential to a city's welfare; it means the decline or growth of a city. I believe this bridge will bring an era of unprecedented prosperity. It will be, in my opinion, the greatest feat of construction ever developed."

After his oration, Strauss answered the War Department's specific concerns. Deakyne inquired as to the likelihood of ships colliding with the piers; Strauss pointed to the 4,000 feet between piers as more than sufficient for the shipping lanes. Deakyne wondered about the effects of an enemy bomb attack, whether the harbor would be bottled up if the bridge were hit. Strauss answered, "If the hit was in the center and both cables were struck, the bridge would fall into 300 feet of water. . . . If it were hit where the cables joined the cantilever, it might block the channel. But then it would only mean a blasting operation to cut loose the other end, and then the whole structure would drop into the bottom of the channel, again opening the way. . . . If the enemy got so close as to be able to bomb the bridge, there would be very little left of the city."

The hearing ended without a syllable of opposition from the floor. Colonel Deakyne, apparently satisfied with what he had heard, assured the audience it would have a decision within two months. It actually took seven, but it was worth the delay. On December 20, 1924, Secretary of War John Weeks notified W. J. Hotchkiss of his department's conditional approval. The terms that Weeks imposed were stiff. The district had to agree "to bear all expense connected with moving, rebuilding, and replacing of elements of the defensive and other military installations damaged by construction; to bear all expense of construction and maintenance of approaches to the bridge; to give the United States complete control over the bridge in time of war; to permit Government traffic at all times free of charge; to provide for wire and pipe lines on the bridge for War Department use free of charge; to subject the construction of the bridge and its approaches, so far as such construction relates to the military defenses of the harbor and the military reservations affected, to the direction of the Secretary of War or his authorized representative." They were taxing but, in the end, admissable concessions.

"The biggest obstacle in our campaign has been overcome," exulted executive committee chairman Hotchkiss. "Now we will start in earnest." Hotchkiss' elation was understandable, but his assessment was wrong. The Coombs Bill and the provisional approval from the Secretary of War were routine steps compared to the steeplechase that was to follow. ■

OBSTACLES: THE LANDOWNERS RESIST

In an age of radio and airplane, the ferry is too slow and uncertain a method of transportation.
—Charles Derleth

Is It Necessary

As soon as the Army had approved provisionally of the cantilever-suspension blueprint, the executive committee launched itself carefully into the north counties. The Bridge District would need tax suzerainty over vast realms of private property, and this could raise widespread opposition.

Initially, Strauss and his people hoped to include ten counties in the District. In the early months of 1925 they stretched themselves thin over the backwaters across the Bay, campaigning upon predictions of economic benefits, and on the promise that the power to tax would be merely a guarantee to indemnify the bond buyers.

Even though the estimated cost of the bridge was rising steadily (up to $21,000,000 in 1925), Strauss clung steadfastly to the formula he had submitted to the Department of War. The District would have to tax, he said, but only to meet the relatively small preliminary costs. Once public funds had paid for feasibility studies and engineering reports, the District would issue bonds that would finance the actual construction. After the bridge was opened for traffic, the revenues from tolls would accrue so handsomely that the District could not only retire the bonds within forty years but also net some $90,000,000 in profits.

On paper, it was a flawless scheme, but in fact it was contingent upon two highly debated points: Strauss's cost estimate, and the Association's projection of traffic. Should either be miscalculated, the taxpayers would be liable to the extent of the mistakes. Many sensed a threat, and fears were left smoldering in the wake of the committee's visits.

Typical was the concern expressed by several north county dairy farmers. They contended the lands on which they earned their livings were actually worth less than they were before the Coombs Bill was passed. The prospect of an expensive bridge, with its implied threat of higher taxation, they claimed, was all that was necessary to decrease property values.

Nonetheless, Strauss and the bridge backers scored substantial early gains. There was an uncomplicated method for joining the District, and the Association expedited it wherever possible. A petition carrying the signatures of 10 per cent of the citizens who voted in the last gubernatorial election obligated a county's board of supervisors to vote on membership in the new District. The first county to yield was Mendocino, north of Marin. It voted itself into the District on January 7, 1925, and was quickly followed by Marin, Sonoma and Napa counties.

Meanwhile, the more reticent areas were watching the progress in San Francisco, clearly the pivotal section of the proposed District. San Francisco not only enclosed the geographical approach to one side of the Gate, it also had the largest population (500,000 according to the 1920 census) and the most extensive tax roll. The city would have 85 per cent of the population of the District. Its property valuation would produce more than three-quarters of the District's tax income. It did nothing to settle insecurities when San Francisco failed to act quickly.

Supervisor Richard Welch had introduced a resolution to include the city in the Bridge and Highway District on January 26, 1925, but the measure was bogged down in committee debate for the next two months. Several committeemen were concerned about San Francisco's disproportionate tax liability; others worried about the limits of its bonded indebtedness. Strauss took time to testify; he told the members that the cost of the bridge would not vary more than 10 per cent from the refigured sum of $21,000,000. Attorney Harlan assured them that financial bonds from the bridge would not affect the city's ability to underwrite other civic projects.

Finally, on March 26, the measure emerged from the committee. But at the Supervisors' meeting that night it was challenged unexpectedly by Acting Mayor Ralph McLeran, who asked instead for $150,000 so the city could authorize a new engineering study of the entire bridge situation.

Welch was enraged at the prospect of beginning anew on matters that had been under consideration since 1919. At the end of a strenuous debate, both measures were postponed a week. When business resumed the following Tuesday, McLeran agreed to withdraw his proposal if San Francisco could have greater representation within the District. He thought that a city with the tax liability of seventeen-twentieths of the district should have more than five-fourteenths of its voting power.

It was a reasonable point and the start of a compromise, but any delay at this juncture was threatening. The entire campaign depended upon a quick and mediating solution. On April 5, north county representatives called a joint session in Napa, where it was decided to rewrite the representation formula in exchange for San Francisco's peaceable entry. An amendment to the enabling act of 1923 was drafted hastily and approved in the State Assembly. It called for one director for counties with populations of 40,000 to 100,000, three directors for counties of 100,000 to 500,000 and a number of directors equal to those appointed from all other counties for areas whose populations exceeded 500,000. Of course, only San Francisco qualified for the latter number. With the reapportionment clause riding the Coombs Bill, Supervisor Welch reintroduced a resolution for the city to join the Bridge and Highway District on April 13. It passed unanimously.

While San Francisco was stumbling through its passage of the Welch resolution, other areas were beginning to defect from Strauss's dream of a ten-county District. Humboldt County, north of Mendocino, refused to enter, citing as its official reason that a Golden Gate bridge was the provincial responsibility of San Francisco and Marin. Unofficially, the motives were a bit more venal. The citizens of Humboldt were largely in favor of a bridge, but the timber industry there feared an infusion of flatlanders and tourists. The lumbermen lobbied diligently against the bridge and were successful in convincing their supervisors to vote it down.

Lake County followed its neighbor's lead. Even though 10 per cent of its eligible voters had signed the Association's petition, Lake County's Board of Supervisors decided 3-2 against joining the district. It was a disturbing trend, this latest opposition, and it turned toward disaster in December of 1925 when Mendocino County, the first member of the District, voted 3-2 to withdraw.

Since Mendocino's entry almost a year earlier, landowners in the county had translated their insecurities into protest. The county grand jury had conducted a hearing and decided that the bridge would jeopardize the singular resources of the county, its land value. Acting on the grand jury's recommendation, the supervisors voted to resign from the district. Then, on January 26,

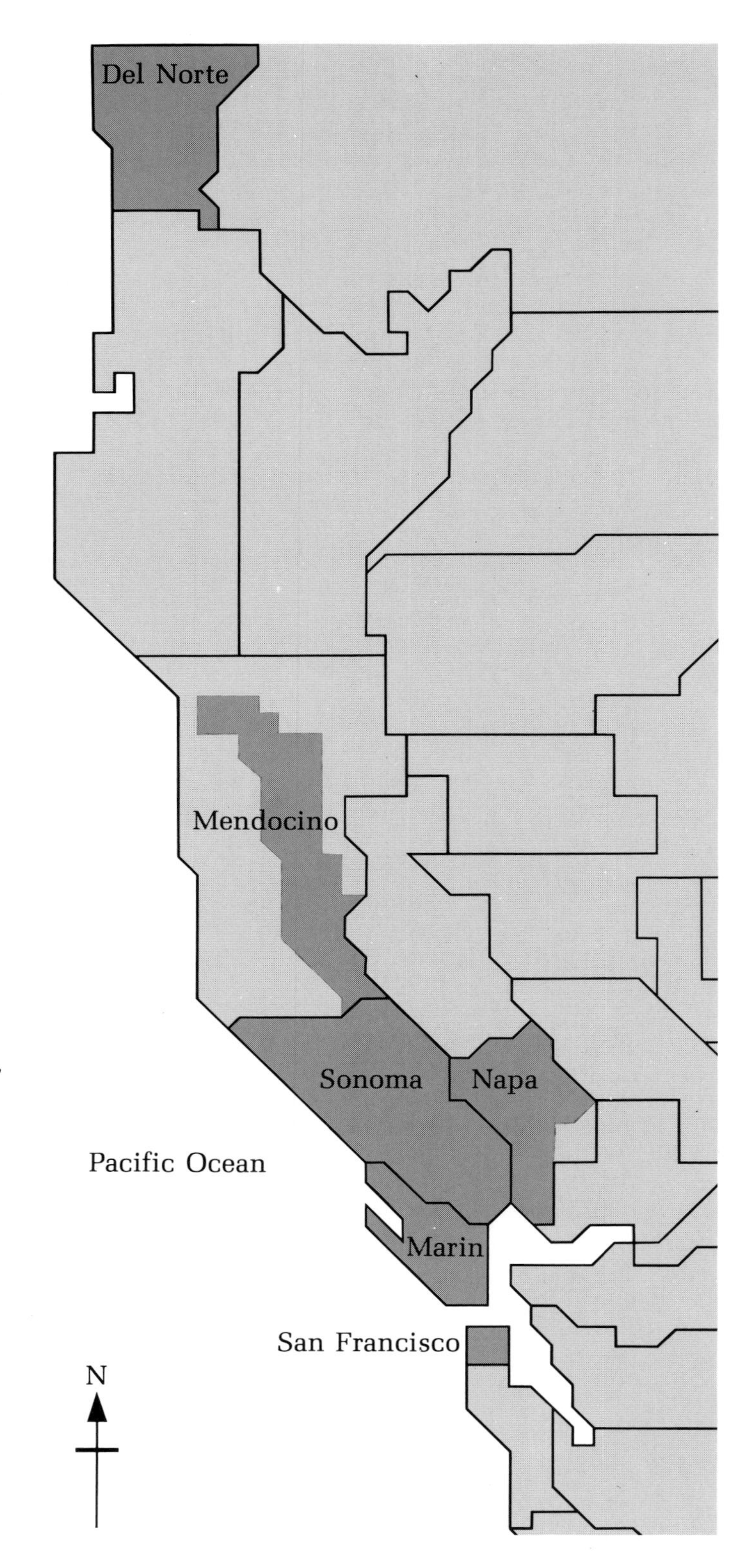

In 1923, bridge backers gained permission from the State Assembly to form a district having the power to levy taxes, issue revenue bonds, and build and maintain a bridge. Originally, the district was planned to include all the coastal counties between San Francisco and the Oregon border. But several counties feared the tax burden and declined to enlist. Then large landowners in two member counties sued successfully to have their properties excluded. Nonetheless, the Golden Gate Bridge and Highway District was legally certified in December of 1928. It comprised San Francisco, Marin, Sonoma, and Del Norte counties, plus portions of Napa and Mendocino counties.

1926, a group of 180 citizens supported the board vote by filing a document which said, in effect, they were withdrawing their names from the original petition the Association had circulated.

The activity in Mendocino echoed all the way to Sacramento, where Secretary of State Frank C. Jordan refused to certify that the bridge District had been organized. He claimed that the 180 signatures on Mendocino's counterpetition brought the number of legal signees down below the necessary 10 per cent. Welch and Doyle, meanwhile, kept a close watch on the impending crisis. When Jordan refused certification, they had George Harlan file suit with the California Supreme Court.

It took some time for the case to be heard, but eventually the high court found for the District. On December 26, 1926, the justices nullified the counterpetition and the second supervisorial vote. With the decision came an automatic order for Jordan to certify the district. The six counties of San Francisco, Marin, Sonoma, Napa, Mendocino, and Del Norte (which had joined without fanfare in August of 1925) became a legal federation subject only to the hearing of individual protests.

The Coombs Bill had given voice to citizens as well as counties. Individual protesters had the right to argue against inclusion of their properties within the District. Secretary Jordan set May 31, 1927 as the final date for submission of protests. By that day some 2,307 cases were on file, 902 from Mendocino, 823 from Napa, 574 from Sonoma, five from Marin, three from San Francisco, and, ironically, none from Del Norte, the county on the Oregon border that would benefit least from a bridge.

After one judge disqualified himself because he was a District taxpayer, the Judicial Council of California selected Judge C. J. Luttrell of neutral Siskiyou County to hear all 2,307 cases. Luttrell opened litigation in October of 1927. In the next six months he would hear complaints from engineers skeptical of a bridge, from timber owners fearing heavy traffic, from farmers panicked over already decreasing land values, from people who thought the bridge would be a blot upon Nature. Nowhere was the reaction more bitter than in the hearings that Judge Luttrell opened in Sonoma in the fall of 1927.

It was there that the north county opposition dug in for a last desperate stand. A group calling itself the Taxpayer's Protective League was organized to represent more than 550 landowners from Sonoma, Napa, and Mendocino. They allied before the bench to argue against Strauss's bridge.

One year earlier, in 1926, the Joint Council of Engineers (which included representatives from the American Society of Civil Engineers; American Society of Mechanical Engineers; American Institute of Mining and Metallurgical Engineers; American Institute of Electrical Engineers; Pacific Association of Consulting Engineers; American Association of Engineers; and the American Chemical Society) had viewed Strauss's design with alarm. The Council was so opposed to the bridge, it conducted an independent study, the findings of which were prime exhibits in the Sonoma hearings.

The men who conducted the Joint Council's research appeared for the taxpayers' group. Under oath, J. B. Pope and W. J. H. Fogelstrom, consulting engineers for the city of San Francisco, and Charles B. Wing, Chairman of the Department of Civil Engineering at Stanford University, assaulted everything about Strauss's plan, including the man's professional integrity, referring to him pejoratively as a visionary and a promoter. They suggested that Strauss's geologists had misread the USCGS channel survey data, that the foundation rock on the south side of the narrows was far too unstable to support the weight of a bridge.

They attacked the design itself, pointing to Strauss's refusal to plan for a suspension bridge as proof he had failed to remain current in his field. They cited recent work on suspension bridges to support their claim—a U.S. Army report which concluded that a 4,300-foot suspension span was structurally possible, and the plans of Gustav Lindenthal, whose George Washington Bridge over the Hudson River in New York included a proposed 3,240-foot main span. Finally, they decried as naive the cost estimates Strauss had compiled.

Comparing his figures with those they had gathered from the Hudson River Bridge, they claimed some startling discrepancies. They used similarities in the materials as their basis for comparisons, testifying that the foundation alone would cost $29,000,000, fully $2,000,000 more than Strauss maintained would be the price of his entire structure. That plus the superstructure and incidental expenses would add up to an astounding $112,000,000.

As an incidental argument, the engineers reintroduced the threat of earthquake damage made imminent by the San Andreas Fault which lay just five miles west of the Narrows of the Gate.

Association attorney George Harlan had Strauss fly in from Chicago to rebut these claims. Predictably, Strauss snorted at the Council's findings, labelling them incomplete and reactionary. From the stand he spoke confidently of the months he had devoted to researching the figures now under attack. The design and estimates, he reiterated, were documented by experts and supported by leading American engineers. He dismissed the doomsaying fear of earthquake damage by calling it a risk all San Franciscans assume.

Following Strauss's testimony, backers from Sonoma took the stand and attested to the economic benefits of closer association with the metropolitan Bay Area. With that, George Harlan rested the District's case.

The Sonoma hearings ended in late November. Judge Luttrell announced he would withhold opinion until he had heard all 2,307 cases. The taxpayers and the Association waited thirteen months, the six it took Judge Luttrell to ride circuit through Sonoma, Napa, Marin, San Francisco, and Mendocino, and the seven more he spent reviewing voluminous transcripts.

In the end, in a ruling signed December 1, 1928, Judge Luttrell disallowed all protests in San Francisco, Marin, and Sonoma, while excluding from the District eighty per cent of Napa and twenty-four per cent of Mendocino. "The cost of construction will not be prohibitive as compared with the revenues reasonably to be expected from the operation of the bridge," he said. "The project is feasible both from the standpoint of an engineering and a financial undertaking."

Judge Luttrell's decision was remanded to the Secretary of State, who then issued the articles of incorporation, declaring the Golden Gate Bridge and Highway District a legal body composed of San Francisco, Marin, Sonoma, and Del Norte counties, plus portions of Mendocino and Napa counties. The District was at last intact. ■

Politics: The Momentum Sputters

San Francisco needs that bridge. We will take the bonds. **—A. P. Giannini**

Railroads and Ferries

Directors for the new Bridge District were chosen by supervisorial appointment. Selections were made quickly except in San Francisco, where city supervisors opened themselves to charges of malfeasance by naming three of their own members as Bridge District directors. The appointments of Frank Havenner, Warren Shannon and William Stanton aroused violent opposition. Editorials criticizing the choices led simultaneously to Havenner's embarrassed resignation and Shannon's and Stanton's staunch refusals to run from what they considered unfair charges.

This was a poorly timed squabble. The vote on bridge financing still lay ahead, and if the first directors were looked upon as quibbling politicians, public sentiment could easily turn against the project. Some members saw the peril. W. J. Hotchkiss said: ". . . the executive committee . . . desired to create and conduct a great public utility free from politics. . . . This idea has been followed in all counties except San Francisco." And from pioneer Franklin Doyle: "One of the chief arguments of the opposition . . . was that if the country counties went into a Bridge District with San Francisco, San Francisco politicians would control and run the bridge. . . . Now the opposition points out its fears were justified."

Recalcitrant Mendocino County took advantage of the appointment crisis to fire one last round at the bridge. Mendocino representatives introduced a bill into the State Assembly which would repeal the 1923 enabling act. George Harlan, as he had done so many times before, went to Sacramento under instructions from the directors to "take such steps as may be deemed necessary."

Harlan spoke so convincingly before the Assembly's Roads and Highways Committee that the Mendocino bill was tabled and replaced by a measure validating all steps taken in the formation of the District. The new bill, drafted by Charles Reindollar of Marin, was passed into law on March 5, 1929. It reaffirmed Assembly support for the bridge and effectively killed off the last of the north county dissent.

While Harlan was fighting for the District in Sacramento, the tempest in San Francisco was dissipating. On January 23, the directors from the city, including Shannon and Stanton, were seated without further opposition. Francis Keesling, who would later chair the District's building committee, was named to replace Havenner. Others on the original board were: William P. Filmer (president), Carl A. Henry and Richard Welch from San Francisco; Robert H. Trumbull (vice president) from Marin; Franklin Doyle, and Joseph A. McMinn from Sonoma; Thomas Maxwell from Napa; A. R. O'Brien from Mendocino; and Henry Westbrook, Jr. from Del Norte. Later, in 1930, when the board would be increased to fourteen members, George T. Cameron, editor of the *San Francisco Chronicle*, and Harry Lutgens, editor of the *San Rafael Independent*, would be added.

By mid-April of 1929 three other positions were filled. Alan McDonald, a local contractor who had built the Mark Hopkins Hotel and the Fox Theater, was named General Manager of the District. W. W. Felt, Jr., for thirty-four years the recorder and clerk of Sonoma County, and a charter member of the bridge association, was appointed Secretary. John Ruckstell of San Francisco was hired as District Auditor. Those appointments left only one significant position to fill, that of Chief Engineer.

Considering his long unsalaried tenure, Joseph Strauss was the leading candidate—but not by anything resembling a wide margin. In his dozen or so years of planning the bridge, Strauss had become something of a liability. His design had been pinned down by critical fire since 1927; his cost estimate was labelled inaccurate at best, irresponsible at worst. His often officious temperament had won him few friends. His maneuvering on behalf of the project made him suspect in many eyes.

In recruiting for the position, the directors considered the candidacies of ten engineers, including Strauss's old boss, Ralph Modjeski. There were also Gustav Lindenthal, whose earlier estimates had disqualified him for the job in 1921; O. H. Amman, Chief Engineer of the New York Port Authority; and David Fowler, designer of the aborted 1915 San Francisco-Oakland crossing. In the summer of 1929 the bridge directors held the equivalent of stage auditions, with all ten candidates flurrying in and out of town, presenting ideas and credentials for review. The decision may have been weighty, but the odds were with Strauss.

Realistically, the job could have gone to no other man. The District owed him something for his tireless efforts since 1919. And Strauss, for his part, was too well acquainted with the District to squander his advantage. Like all candidates, he submitted a written proposal.

In it, he appealed to sentiment, citing the long and loyal years he had virtually donated to the bridge movement. He addressed tax fears by guaranteeing a cost he knew the property owners could stand. He allayed the doubts of the Joint Council by promising to hire O. H. Amman and Leon Moisseiff as consulting engineers. Both men were principals in the George Washington Bridge, the span the Joint Council had used as a basis for its computations, the one bridge that clearly had the Joint Council's unqualified support.

And if his proposal were not convincing enough, Strauss had the help of a companion that summer, a Dr. Meyers from Los Angeles, who was reported to be his advance man. It was odd to see the friendless Prussian engineer shadowed by a glib and polished doctor with no apparent practice. Meyers was a mannered gentleman with a gift for congeniality. He threw revelling parties in his suite at the Palace Hotel that summer, to which he invited both politicians and influential members of the Bridge District hierarchy.

Whatever the reasons for the directors' choice, Strauss was named Chief Engineer in mid-August of 1929. As promised he hired both Amman and Moisseiff as consulting engineers. For a third position, he enlisted Charles Derleth of the University of California's College of Civil Engineering. Derleth, mastermind of the Carquinez Bridge, was himself a candidate for the job of Chief Engineer of the Golden Gate and, most significantly, was a vocal critic of Strauss's cantilever design. It was a staff created for an eclectic approach to spanning the Gate.

Then, in a move probably more important than any other, Strauss abandoned his cantilever-suspension hybrid. His reason was partly political; he felt that the bridge simply could stand no more public suspicion. But Strauss had also become a belated convert to the suspension bridge. Advances in metallurgy had made a center span of more than 4,000 feet technologically possible, and work by Leon Moisseiff had disproved the need for a rigid bridge to face howling winds. In collaboration with Frederick Leinhard, Moisseiff had concluded

In the summer of 1929, with the concurrence of his newly appointed staff, Joseph Strauss abandoned all ideas for a cantilever-suspension span. In the next six months Strauss's assistant for design and construction, Clifford Paine, developed a blueprint for an all-suspension bridge. The result was submitted in the spring of 1930 and became the official rendering of the Golden Gate Bridge. The long-awaited project was now ready to begin, subject only to authorization from the military, the financiers, and the voting public.

that the cables of a suspension span could deflect up to half the impact the winds could impose. Further, a bridge flexible enough to bend with the pressure would suffer far less structural weakening.

In all his accounts Strauss assumed full credit for the change, but it was actually sponsored by a junior partner in the Strauss firm, Clifford E. Paine. Paine was a brilliant engineer whom Strauss brought west to supervise design and construction of the Golden Gate Bridge. Throughout the project, Paine would make most of the major decisions.

"Oh yeah, Paine built the bridge," says Ted Huggins, a retired public relations man from Standard Oil who was on loan to the Bridge District for the duration of the Golden Gate project, and a friend of Strauss's dating back to the 1920's. "You see, it was like a corporation, where the chairman of the board makes policy decisions, supervises planning and assumes the responsibility. That was Strauss. But Paine was the man who oversaw the day-by-day operation."

Throughout their association Paine and Strauss maintained a productive if not always harmonious relationship. "They fought like alley cats all the time," remembers Ruth Natusch, "but I don't think one could have gotten along without the other." Indeed, theirs was a rare bond. Paine was inspired by Strauss's imagination, but subdued enough to remain contentedly backstage. In turn, Strauss's often wild ideas were held practical by Paine's unyielding good sense.

Paine was the only member of Strauss's firm unintimidated by the boss, and the two would argue a concept to a satisfactory compromise. Presumably the decision to switch to an all-suspension bridge rose from such discussions. In any case, it was adopted unanimously by the staff at its very first meeting, held in August of 1929, less than one week after Strauss was named Chief Engineer for the Golden Gate Bridge and Highway District.

In the twelve years since Joseph Strauss was first approached on the Golden Gate, the project had moved forward. But now even after a dozen years of theorizing, legislating, and politicizing, the bridge was still far from a reality. The strength of the bedrock of the south pier had yet to be determined. New blueprints accommodating the changes in design had to be drawn up. They in turn would have to be sent to the War Department in Washington for final approval. And, of course, there was still the matter of finances.

Before actual construction could begin, the public would have to market the legal bonds. This last may have seemed a mere formality in the summer of 1929, with the Bridge District sanctioned and mobilizing for a breaking of ground. But three months later a crashing stock market would plunge the United States into her worst depression ever, taking with it any grounds for District optimism.

The District invited bids for foundation borings in September of 1929. After consideration, the contract was awarded to the E. J. Longyear Exploration Company of Minneapolis. Analytical work proceeded that fall, with Strauss's newly hired geologist, Andrew C. Lawson of the University of California, the man who had discovered and named the San Andreas Fault, assaying the data along with his collaborating associate, Allen Sedgewick of the University of Southern California. Strauss and his staff were pleased with their reports that the bedrock near Fort Point, while probably not ideal, would suffice. "The tensile strength of this rock is low," Lawson's findings read in part, "but it is entirely adequate to support the terminal pier of the bridge."

By February of 1930 the foundation borings and engineering studies had been translated into a formal report which Strauss submitted to the directors. In it, he accounted for the conversion to an all-suspension bridge and other, more subtle changes.

From the north end soundings it was determined that the Marin pier should be moved shoreward by 200 feet, both to minimize costly underwater construction and to take advantage of a subaqueous ledge jutting out from the shoreline. It was a necessary decision but it meant the already unprecedented length of the center span would now reach 4,200 feet between the towers.

For more structural strength, and at Moisseiff's suggestion, the roadway was widened from the 80 feet in the original design to 90 feet. The staff, reasoning that the age of interurban railways was passing, also eliminated the center tracks that had been a conspicuous feature of Strauss's cantilever bridge. Those two alterations would allow six 10-foot lanes of traffic, a number so great by the standards of the day that even the visionary Strauss believed the bridge would never know a traffic snarl.

Some of Strauss's more elaborate ideas were also jettisoned. Cast aside were plans for a masonry toll plaza that would have resembled the arch of triumph and a set of gilded wrought iron gates hung between the bridgehead pylons bearing the inscription "Golden Gate Bridge." Gone were ideas for a glassed-in elevator that would have taken visitors to an enclosed tower-top observation deck. Also abandoned was Strauss's Byzantine scheme for supporting the roadway over Fort Scott by slinging a cable underneath the deck. At Paine's insistence, the fort would be spanned by a conventional steel arch.

The directors were delighted by the engineering report of 1930, placated at last about the strength of the bedrock and the principles of design. They were hopeful now of an unobstructed bond campaign, with only the formality of War Department approval in the interim. That spring Strauss and his engineers finished the preliminary drawings and sent them off to Washington. Unexpectedly, the Secretary of War announced that his department would have to conduct an entirely new hearing, citing the time elapsed since 1924 and the drastic revisions in the 1930 design.

News of the War Department's intention sent District president Filmer and attorney Harlan to Washington, where they enlisted the aid of Richard Welch (then a congressman) and California senators Hiram Johnson and Samuel Shortridge. With these heavyweight allies, the District representatives persuaded the War Department to rescind its latest order. It was a short-term victory.

Shortly after Harlan and Filmer returned to the Bay Area, local shipping interests protested officially that the new plans failed to allow sufficient clearance for highmasted vessels. Once more the War Department reversed itself, this time with appeal. There would be a new hearing, to be convened the first week of July, 1930. In something of a concession, it agreed to consider only the issues of vertical clearance for navigation.

Major General Lytle Brown presided over the 1930 hearing, held again in the board chambers of San Francisco City Hall. As he did six years earlier,

Joseph Strauss served as bridge advocate. The issue of clearance was obviousiy critical, but it was dispatched by a sarcastic exchange between Strauss and a shipping company attorney who was questioning his calm assertion that the distance between the bridge and the water could vary 16 feet.

"You mean," asked the attorney, "that the clearance will be sixteen feet greater at low tide than at high tide?"

"No," replied Strauss. "What I mean is that the cables will lengthen on hot days and lower the bridge by as much as sixteen feet."

"Oh, so you are building a rubber bridge."

Strauss's offhanded explanations apparently satisfied Brown. In a ruling dated August 11, 1930, the War Department issued its permit for a 4,200-foot span with a mid-length clearance of 220 feet and a sidespan clearance of 210. General Brown said that he was opposed personally to the bridge, but that he was mandated by the 1924 order to approve the permit.

Cleared by the Army, the staff proceeded with the final drawings. Strauss had hoped to complete the design and open construction bids by January of 1931, but as had been the pattern since 1917, the project was derailed once more.

By the summer of 1929, a tax of 3 cents per $100 had already been levied against District properties; a year later an additional 2 cents was added. In all, $465,000 had been raised. Taxation was intended only until the financing bonds could be sold. And now, in the middle of the Depression, the District had to sell the taxpayers on the idea of a $35,000,000 bonded indebtedness.

The engineering report of 1930 had eliminated most doubts about the bridge, but the directors knew that an affirmative bond vote would not follow automatically. In the fall of 1930 they released $50,000 from a public relations war chest and brought their message to the public through radio spots, newspaper ads and public speeches encouraging a "yes" vote on Proposition 37.

All of San Francisco's daily papers came out in support of the bridge, with editorials ranging from the almost blind enthusiasm of the *Examiner*, ". . . the mightiest suspension bridge in the world will soon span the Golden Gate!", to the resigned approval of the city's largest paper, the *Chronicle*: ". . . the bridge when built will be worth so much to San Francisco and its whole district that it will be worth even a considerable deficit for some time. . . ."

The theme of economics was reiterated time after time. In countless addresses District general manager MacDonald claimed that "the Golden Gate Bridge will cost the taxpayer his vote, and that is all." And the president of San Francisco's Chamber of Commerce announced himself "tired of questions on how to solve the unemployment problem. . . . It's the job of every voter in the city to create jobs by voting for the bond issue."

Opposition was predictable, widespread, organized, and sometimes venomous. Shippers claimed a bridge would cripple the Bay Area's vital maritime industry. A taxpayers' committee once again brought up the familiar demons, the puddingstone rock under the south pier site, and the $112,000,000 estimate of the Joint Council. Even M. M. O'Shaughnessy, once the foster parent of the span, publicly questioned the building of a bridge during the Depression. He had seen the cost of his Hetch-Hetchy project multiply unexpectedly during the sixteen years it had been under construction, and he feared the same with the Golden Gate bridge. And, of course, there were scattered but vehement letters to the editor and public speeches that decried the bridge as an unspeakable mar on the Golden Gate.

But in the end the protests gave way to a Depression backlash. Uppermost in the minds of the impoverished electorate was the fact that $35,000,000 in construction would translate into about $750,000 a year in wages. Proposition 37 carried overwhelmingly, by a 3-1 average margin, and by as much as 8-1 in Marin and Sonoma.

After the election, Strauss and his staff went to work on the final design and specifications, completing them in April of 1931. The District invited construction bids that same month. By the time the bids were to be opened—in mid-June—twenty-seven firms had offered forty proposals in competition for the ten contracts. The total of the low bids, the directors discovered at their June 17 meeting, amounted to $24,955,299, almost $3,000,000 less than the estimate presented to the voters for the 1930 election, and final, graphic testimony that Strauss's calculations had been accurate all along. Provisional contracts based on the lowest bids were awarded to ten firms that summer of 1931, the very point at which the now rolling project stumbled and fell.

A committee charged with the responsibility for preparing marketable bonds had been appointed immediately after the election. It in turn had hired specialists from both the east and west coasts to assist in the creation of policies and procedures. The committee initially decided to offer a block of $6,000,000 worth of bonds for sale, with smaller units to follow as needed. It was a solid portfolio for a bullish market, but one powerless to quiet Depression anxieties. Hard times had retired all but the most cautious investors, and demand for Golden Gate bridge bonds receded with every passing rumor. Whispering doubts over the bedrock of the south pier, over the Joint Council's $112,000,000 estimate, over the constitutionality of the District's right to collect taxes had made the Golden Gate bridge an unclean commodity in a bond market turned fastidious by the Depression.

To loosen investor conservatism the committee recommended a court test that would finally and undeniably remove any doubt as to the District's right to guarantee the bonds by the taxation of private property. The directors vetoed the suggestion at first, apprehensive of another long court battle. Determined to avoid any more judicial wrangling, the District offered a block of $6,000,000 worth of bonds for sale on July 8, 1931. Only one offer was tendered, and it was conditional. Dean Witter & Co. was willing to buy the bonds, but only if a court test first determined the legality of the taxing powers, an independent study conducted by engineers of Witter's choosing verified the solidity of the bedrock near Fort Point, and it was agreed that Witter could cancel the offer without penalty any time up to the delivery of the bonds. The Directors found Witter's bid unacceptable; they denied it after cursory debate.

Several days later another, more reasonable offer was received from a syndicate headed by BankAmerica Corporation. It would buy all the bonds at an acceptable rate; it would impose none of the conditions of Witter's bid—except the court test. Sensing the inevitable, the bench-weary directors decided at their July 16 meeting to enter court once more. They arranged for District Secretary Felt to refuse endorsement for the bonds, whereupon they filed suit against him.

The District had hoped to settle the friendly suit against Felt by September of 1931, before the provisional construction contracts lapsed and while the BankAmerica bond offer was still valid. Quick adjudication, however, depended upon a dormant opposition. Unfortunately, The Golden Gate Bridge and Highway District vs. Felt caused controversy to rage once more. A law firm representing "anonymous taxpayers" entered petitions in the hearings. Attorney Warren Olney charged that the Bridge District had been illegally formed, that the number of signatures on San Francisco's petition for membership fell below the requisite 10 per cent of eligible voters (which, owing to a miscalculation in the office of the Registrar of Voters, was in fact true), and that as a result, 85 per cent of the District's population and more than 75 per cent of its assessed valuation was illegitimate.

The District was unmoved either by the merits of the claim or this latest search for a flaw. But its members did worry about the source of the opposition. The bridge project was close to bankruptcy; even short delays were of urgent concern. The trial had frozen the only certain source of income, the bond payments from BankAmerica. Strauss was maintaining a full and salaried staff of idle engineers. Expenses were increasing at a rate of $4,400 every thirty days. If the litigants were drawing from bottomless accounts, they could prolong the battle until the Golden Gate bridge project was torn asunder—regardless of the outcome in court.

The identity of the protesters remained undisclosed until September 15, when Olney admitted that he represented ninety-one taxpayers and the Southern Pacific-Golden Gate Ferries Ltd., the company whose monopoly on north bay crossings would automatically dissolve with the opening of a bridge, and whose majority owner (51 per cent) was the vast, influential Southern Pacific Railroad.

The ferry company claimed only passive interest in the lawsuit, saying that it was no more or less involved than any of the ninety-one citizens who regarded the Golden Gate bridge as injudicious. For its part, Southern Pacific denied any connection whatever. Both claims were met with howling disbelief. Bridge backers refused to accept that SP would allow a minority company to dictate policy; they refused to accept that the ferry company was so touched by public spirit that it was merely trying to prevent the Bay Area from a costly error. The directors assumed they were at war with vested interests of considerable resources and leverage.

The trial delayed the bridge project by several months. The bond offer from BankAmerica, carrying a 120-day time limit, remained live only until the 16th of November, 1931. Ferry company petitions extended the hearing nine days past that deadline. On November 25, the high court ruled 8-1 for the bridge. It was a moral victory, but the directors had to return BankAmerica's $120,000 performance guarantee and readvertise the bonds. The power to tax was once more validated, but the District was no better off than it had been on July 8; in terms of escalating expenses, the situation was substantially worse. What's more, there was every indication that the ferry company would continue to sabotage the project.

Three days after the Supreme Court's ruling, the Garland Company, north county property holders acting in the interest of Southern Pacific-Golden Gate Ferries Ltd., filed suit with the Federal District Court of Appeals in San Francisco, asking that the bridge directors be enjoined from selling bonds. Grounds for the appeal lay again in the taxing authority. The suit claimed that the landowners in six counties had been assessed in violation of due process.

In Depression-wracked 1931, the bridge was fortunate to have a foe so redolent of big business. During the sixteen months since the bond election the bridge had become almost mythic as a symbol of unemployment relief. The public fumed at the idea of a corporate titan like Southern Pacific standing in the way of new jobs.

All the hostility the District and its allies could muster was fired at its adversaries. Newspapers castigated the ferry interests, accusing them of sacrificing the common good for their own gain. The Motor Car Dealers Association of San Francisco passed a resolution condemning Southern Pacific and called for a boycott against rail freight. Civic groups in San Francisco and Marin contacted local, state, and national businesses, asking for cooperation in the boycott. A citizens' group organized for the purpose of intensifying the boycott elected as its chairman Frank McDonald, president of the San Francisco Labor Council, implicitly throwing the weight of organized labor behind the bridge.

The group brought its message directly to the people, its spokesmen denouncing Southern Pacific and the ferry company on nightly radio and in the lecture halls. Various service and fraternal clubs placed themselves on public record against the bridge opponents. Even the city of San Francisco took an active role. The Board of Supervisors passed a resolution demanding that the ferry company spend several hundred thousand dollars repairing and enlarging its Hyde Street Ferry slip or relocate the slip altogether.

The pressure began to break down the ferry company's resistance. After first threatening to prolong litigation, ferry company president S. P. Eastman agreed to help expedite the trial if the board would relax its demand for slip improvements. In another concession, Eastman offered to drop the suit altogether if the bridge could be constructed and operated by the recently formed California Toll Bridge Authority (as was the Bay Bridge) instead of by a separate District underwritten by property taxes.

But it was too late for compromise. After the Southern Pacific and the ferry company had admitted their roles in the current litigation public resentment had turned acrid. This latest offer only redoubled the bitterness. By the time the trial reached the chambers of Federal Judge Frank H. Kerrigan on February 16, 1932, SP and the ferry company were yoked in public contempt.

The matter remained unsettled for five more months. The actual trial required little more than a week, but Judge Kerrigan weighed opposing claims all spring. In July he found for the District, ruling nothing improper about either the methods for organization or the provisions for taxation as granted by the Coombs Bill. Immediately the ferry company announced it would appeal, all the way to the Supreme Court if necessary, despite public pleas by both the San Francisco Board of Supervisors and the Grand Jury.

Further court wars could delay the bridge interminably, possibly past the point of ever being recouped. This time, however, the fears were short-lived. Early in August, Eastman stated publicly that his firm could no longer stand the assumption that its only interest was to monopolize north bay crossings. The bridge, he reiterated, was ill-timed, ill-conceived, underfinanced, and unconscionable in its restriction of taxpayer rights. In spite of all that, he said, Southern Pacific-Golden Gate Ferries Ltd. would no longer oppose its construction.

Southern Pacific retired to its corner, but it was the Bridge District that was financially exhausted. It had assessed taxes twice and was still in debt for more than $250,000. Returning open-palmed to the taxpayers could alienate public opinion. Appeals had been made to the Reconstruction Finance Corporation, but they were tangled in paperwork and would ultimately be rejected. The crisis was aggravated by a weak response to the readvertised block of $6,000,000 worth of bonds.

Of the sixty houses invited to make bids, only two replied. Of those, only one met the stringent requirements of the District. This was another offer by the BankAmerica Corporation, which would buy all $6,000,000 worth and immediately advance the depleted project $200,000. The promise of a fast injection of funds was elating, but scrutiny by bond specialists in New York (the west coast specialists had declared the offer satisfactory) found the bid invalid. Its interest yield surpassed the ceiling of 5 per cent that had been established by the Coombs Bill in 1923.

After all the legal battles, won each time by the District; after all the public information which successfully convinced the electorate of the need for a bridge; after the wide plurality in the election of 1930; after staring down the threats posed by the monolithic Southern Pacific, the beleaguered Golden Gate Bridge project was now subverted by a Depression-sapped bond market which yielded but one legitimate offer, one that was disqualified by a technicality in some ten-year-old legislation. Clearly, the bridge needed a benevolent turn of fate.

All through the Depression the head of the Bank of America, Amadeus P. Giannini, had seen his company as a vanguard of recovery. To facilitate the swing back from hard times, he had instituted a program of incentives, reinvestment in the courage, the spunk, the ingenuity of the American people. It amounted to fiscal jingoism, but Giannini's bank was committed to more than just slogans. It was far less conservative than most bond houses when it came to local purchases.

When Giannini learned the Golden Gate bridge was sinking fast into the red (he found out because Strauss and a delegation from the District approached him on the subject), he saw immediately the civic necessity of funding the project. "I know," He is reported to have said to the more conservative members of his board, "San Francisco needs that bridge. We will take the bonds." With Bank of America president Will Moorish, Giannini conceived of a novel plan.

The District would accept the BankAmerica Corporation's offer to buy $6,000,000 worth of securities at the higher rate. While the legality of that purchase was being tested in court, the bank would immediately buy up $3,000,000 worth of bonds at the approved 5 per cent rate. That second purchase could be cancelled, according to the agreement, if the original $6,000,000 bid were declared legal, making the higher rate the operating figure.

The Bank of America's revised bond purchase was accepted by the directors, ending the District's most severe financial crisis. But there were still some shaky times ahead. February 1933 proved to be the worst month for banking in the history of American finance. B of A was forced to close its doors on March 4, threatening the District's solvency. But Giannini convinced federal authorities of his firm's solidity, and he was allowed to reopen on March 13. Then in April the courts agreed that the District could issue bonds which yielded a return greater than 5 per cent, and the cash flow was even more firmly insured.

The fresh capital from B of A allowed construction to begin in January of 1933. From then on, the bridge project would roll without break until it was completed in May of 1937.

It had been a long, exhausting, expensive and, at times, dirty fight. Joseph Strauss had been involved with every obstacle along the way and he had helped clear them all—but not without cost to himself. In that decade and a half, he was slowly drained, psychologically, emotionally, financially, and for a short time, spiritually.

As construction began, he indulged in the well-deserved luxury of a breakdown, spending the first six months of the new year recovering in the Adirondacks while Clifford E. Paine was building the bridge without him.■

Groundbreaking: The Work Begins

Man has upset restrictive precedents, exploded prejudices, and has apparently defied the very laws of nature in setting forth on this great adventure.
—San Francisco Chronicle

A Civic Holiday

In the early summer of 1932, with insolvency encircling the bridge project, the District released all construction firms but one from their provisional bids. The future of the bridge was far too murky to remain contractually tied to nearly $25,000,000 worth of building promises. Not until the surrender of ferry interests in August of 1932 and the transfusion of Bank of America money two months later could the directors securely reopen the bidding. When they did, they found themselves in a position more favorable than the one they had relinquished.

All companies under provisional contract would have been happy to resubmit their earlier bids; in the middle of the Depression, none would have risked the loss of so immense an assignment. But both Strauss and the head of the District's building committee, Francis Keesling, believed that men and materials would be cheaper than they had been a year and a half earlier. Except for the largest of the contracts, issued to McClintic-Marshall for the supplying and building of the 75-million-ton steel superstructure, the directors asked for all new bids. They retained McClintic-Marshall at its original bid of $10,494,000 (a figure Strauss and the building committee had always considered reasonable) because it operated an assembly plant in the Hunter's Point section of San Francisco, and the District wanted wherever possible to stimulate local industry.

Of the approximately $24,000,000 in contracts awarded at the November 4 Board meeting, about $7,500,000 went to west coast industry.

A cable spinning contract amounting to $5,855,000 was granted to the New Jersey firm of John Roebling & Sons, but the remaining awards were regional: The Pacific Bridge Company would be paid on two contracts, $2,935,000 for constructing the piers on both the San Francisco and Marin sides, and $555,000 for pouring the roadway across the bridge. Work on the steel superstructure of the San Francisco and Marin approaches would be shared by J. H. Pomeroy & Company and Raymond Concrete Pile Company for $934,000. Eaton & Smith would build San Francisco's approach road for $994,000; Chigris and Sutsos, Marin's for $59,780. Barrett & Hilp was granted a contract of $1,859,855 for constructing the anchorage houses on either side, and for building the minor piers and cable housings. All the electrical work would be installed by

Alta Electric & Mechanical Company for $154,000.

More remarkable was the final total of the contracts, $23,843,905, or $459,000 less than the 1931 figures. Massive bridges less challenging—notably the Philadelphia-Camden and the Hudson River bridges—would cost millions more. But in those spans the prices included the purchase of vast tracts of urban real estate with associate expenditures of razing buildings and clearing the rubble. Fortunately, all lands approaching the Golden Gate were undeveloped easements donated by the military. In the end, Strauss would build a bridge rivalling the most awesome ever conceived, for less money than it takes nowadays to construct and equip a 747 jet airplane.

Preliminary work on the bridge began in late November of 1932, when a 1,700-foot access road was cleared from Waldo Grade down the hill to Lime Point. Actual construction started without ceremony on January 5, 1933, with two gigantic steam shovels opening excavation for the foundation pit of the Marin anchorage. Ground was not officially broken until February 26, a clear chill day in which even the winter fog saluted the bridge, rolling in only as far as the western hills then hanging there for the entire afternoon.

The ground-breaking ceremony might look excessive, even chauvinsitic, through a backward glance of so many decades; it was a celebration of the bridge's popularity and its symbolic identity as a monument to progress and relief from hard times.

At 12:45 an artillery barrage began a parade. Army, Navy, Marines, veterans, American Legion, fraternal organizations, National Guard, and Boy Scouts passed in review before excited throngs lining the streets and dotting porches and rooftops of the Marina District. They were met by a huge crowd at Crissy Field. Overhead, smoke painters brushed a billowy image of a bridge across the high blue sky. Navy planes flew a formation in salute. A group of engineering students from the classes of Charles Derleth at UC Berkeley strode in carrying an 80-foot replica of the Golden Gate Bridge.

The crowd was so lovingly boisterous as to constitute a menace. The officials had to pull and shove just to clear a space for the actual groundbreaking, which followed platform speeches by Governor James Rolph, Mayor Angelo Rossi, and District President William Filmer, and the reading of a congratulatory telegram from President Herbert Hoover.

At 4:00 PM the dignitaries descended to the open ground. There, Major General Malin Craig handed official rights-of-way grants to William Filmer, whereupon Rossi and Filmer took up a golden spade and jointly drove it into the ground, overturning clods of earth to begin official construction of the Golden Gate Bridge.

The spade was handed to Strauss for another symbolic turn of ground, but there the ceremony was halted prematurely by Supervisor Shannon, who feared the rumbling crowd was about to erupt. It had already pushed threateningly against the platform, damaging the bridge model. The huge assembly roared with lusty good nature at its dismissal, feeling not the least cheated by its failure to witness the setting of a commemorative plaque, or the planting of some memorial redwood trees. Happy at just being part of history, and elated by the sound of its own cheering, it dispersed without rancor.

From the day of the groundbreaking forward, the tenor of the bridge project changed. There would still be political intrigue; the bridge did after all represent the kinds of wealth and power which attract the high-rollers of finance. BankAmerica Corporation, the fiscal hero of the prehistory, had already used its leverage in January of 1933 to pressure General Manager McDonald into resigning (objecting, as had some directors, to his association with a backstage political figure named Abraham "Murphy" Hirshberg). Public doubts over the bedrock of the south pier would threaten again in the spring of 1934. One director would resign amid hints of collusion over a large order of silica cement sold to the bridge by a plant in Santa Cruz that he (the director) owned.

But those were incidents floating in the breeze over a bumper harvest of vision and purpose. Though the men in vested pinstripe suits and smartly blocked fedoras would still play their parts, they were in the shadows now, instrumental but peripheral. From the day of the groundbreaking forward, the Golden Gate Bridge would belong to the men in hard hats and brown canvas overalls, men who would carry lunch buckets and work all day shivering in oilskin parkas at dizzying heights over treacherous waters.■

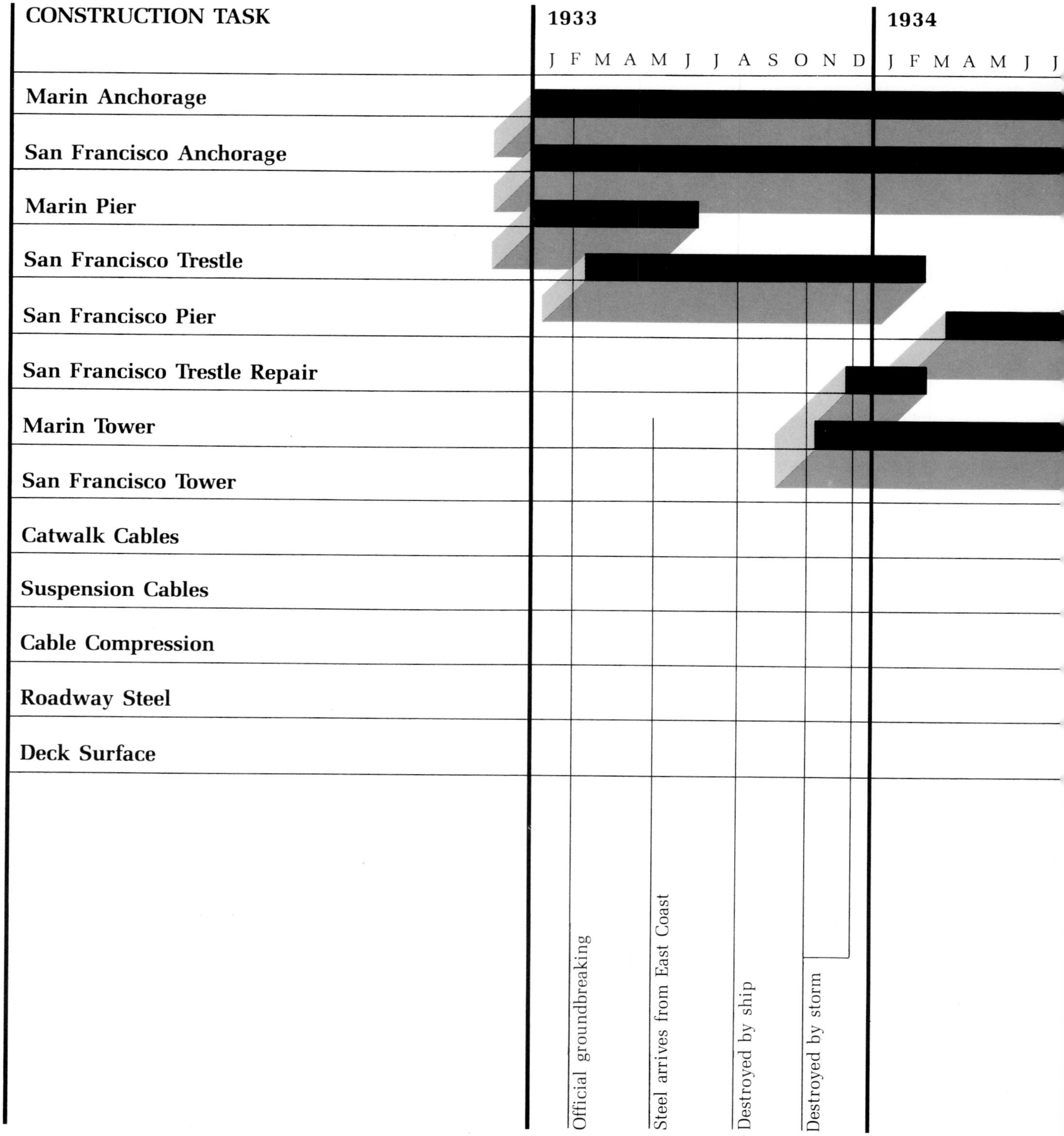

CONSTRUCTION TASK	1933 J F M A M J J A S O N D	1934 J F M A M J J
Marin Anchorage		
San Francisco Anchorage		
Marin Pier		
San Francisco Trestle		
San Francisco Pier		
San Francisco Trestle Repair		
Marin Tower		
San Francisco Tower		
Catwalk Cables		
Suspension Cables		
Cable Compression		
Roadway Steel		
Deck Surface		

Official groundbreaking

Steel arrives from East Coast

Destroyed by ship

Destroyed by storm

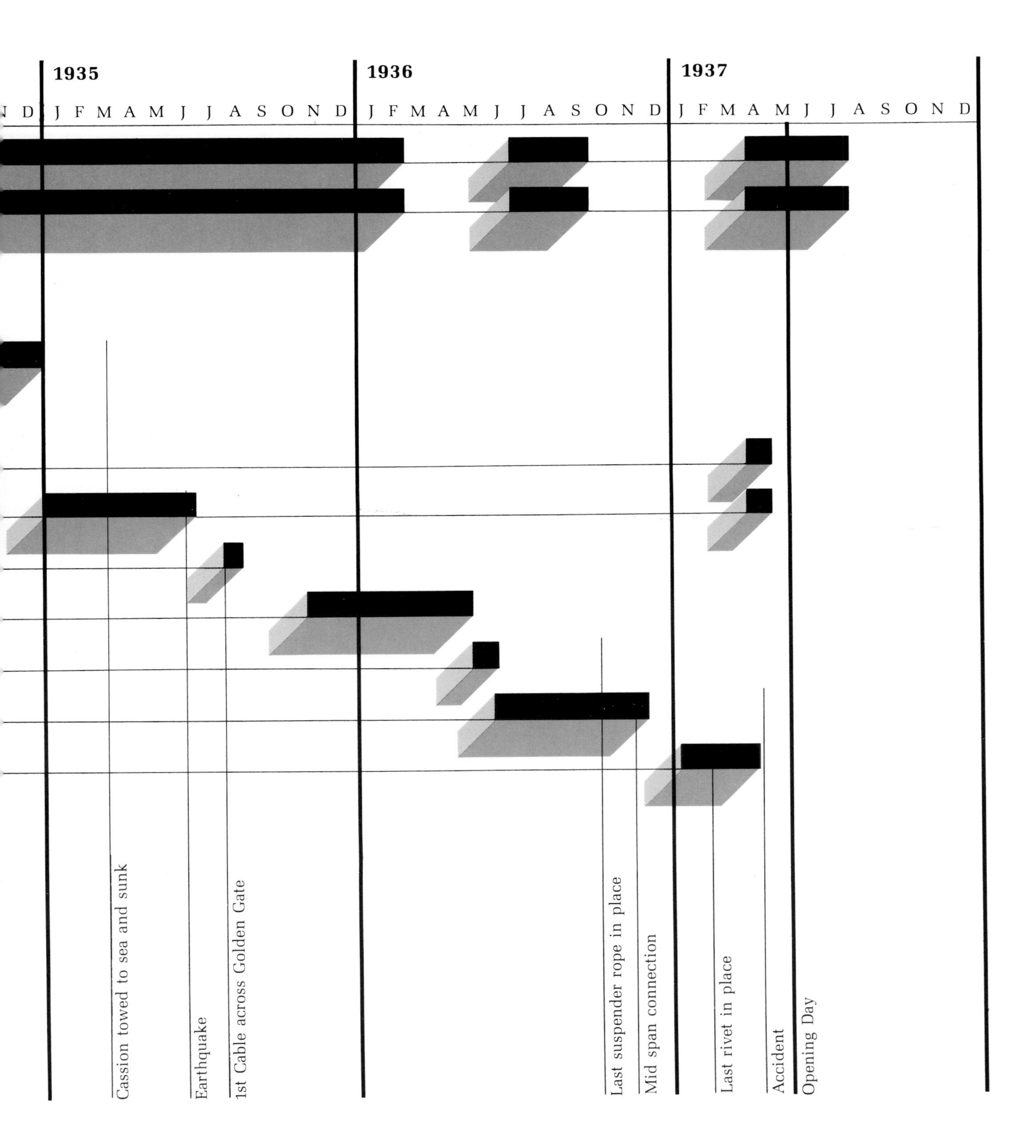
1935
1936
1937
N D J F M A M J J A S O N D J F M A M J J A S O N D J F M A M J J A S O N D
Cassion towed to sea and sunk
Earthquake
1st Cable across Golden Gate
Last suspender rope in place
Mid span connection
Last rivet in place
Accident
Opening Day

FOUNDATIONS: THE SEA RELENTS

The cement required for the piers and anchorages of the Golden Gate Bridge would make enough concrete to build a five foot sidewalk from New York to San Francisco.

Supporting a Bridge

The Golden Gate Bridge and Highway District felt strongly obliged to help disperse the breadlines in the Bay Area. Public opinion had supported the project partly because it stood for unemployment relief, and organized labor was influential in the victorious bond election of 1930. As a consequence, the District had established labor-approved guidelines as early as the summer of 1930. And now two and a half years later the District reaffirmed its earlier vows.

The Golden Gate Bridge was to be, where possible, union-built by District residents (at least one year) working for respectable wages ranging from $4 a day for unskilled labor to $11 a day for skilled. This was an admirable position, progressive and fair for the 1930's when many unions were still a decade away from negotiating coequally with management.

Some of the contractors still scorned organized labor, but most stood somewhere between the attitude of Pacific Bridge, which hired union and nonunion alike, allowing the unions to solicit membership on the site, and that of J. A. Roebling & Sons, which demanded union membership of all its workers.

The residency rules were violated only sporadically. The District allowed for a certain percentage of "key men," supervisors and specialists who had to be imported, and their numbers stretched beyond what was intended. Also, many workers with southern drawls and East Coast inflections found their way onto the project with addresses bribed from innkeepers, or with doctored utility bills dating back the necessary twelve months.

Still, these were the exceptions, and on balance the Bridge was responsive to the plight of the locally unemployed. The construction site was no place for the indolent or the weak, but those hungry men who could endure both fear and fatigue at least had the chance for a job.

There were times during the Depression that my wife and I wouldn't eat just so the baby would have enough food. I remember some days I walked along Market Street from the Ferry Building to Van Ness Avenue looking for a job. It might take a half day to walk up one side of the street and a half day to walk down the other. I'd knock on the door of every business on that goddamned street, and, like as not, I'd get turned down at every one of them.

Then the Bridge came along, and some of us found work. I remember going out there one day—they hired and fired on the spot—and talking to one of the Pacific Bridge foremen, Blackie Silver. I said to him, "How about a job?" He looked me over for a second, then asked, "You ever worked high?" I said, "Mister, I'm looking for a job. It don't matter where or what. I need work." So he pointed out toward the roadway—this was when they had just finished putting up the steel across the Gate, and there was nothing out there but girders. He said, "See those panels out there?" I said, "Yeah, I see them." He said, "Go get 'em."

His words hit me right in the pit of my stomach. The panels were like 100 feet out toward the middle. That meant walking along those girders with nothing to hold on to, balancing myself on 8-inch I-beams with only net and water underneath. The thought of walking the flanges scared the hell out of me. But I did it. I learned quick that when the wind was blowing, which was all the time out there, you had to carry lumber on the side away from it. If you didn't, it could get hold of you and blow you right into the drink.

I don't know how, but I managed to make it back all in one piece. I must have been whiter than a sheet. Blackie looked at me a while. He said, "I see that you're scared." I said, "You're goddamned right I'm scared." He hesitated for a second. "Well at least you got guts enough to go out there . . . Okay, you can go to work." (Pete Williamson)

A suspension bridge is a marvel of physics, a massive structure of interdepending parts, all sharing the stresses imposed by the span's enormous weight. For the Golden Gate, 389,000 cubic yards of concrete, 83,000 tons of structural steel, and 24,500 tons of bridge wire were assembled and converted into the foundations, the towers, the cables, the floor, and the roadway.

As with any suspension bridge, the Golden Gate was to follow a strict schedule of construction. Interest on the bonds was accruing at nearly $5,000 a day, demanding from Strauss, his staff, and the contractors the greatest efficiency, maximum speed within reasonable cost.

In the final months of 1932, between the times the bids were accepted and the final contracts drawn, Strauss and Paine examined the logistics of the task that lay ahead. They allowed for differing composition in the ground that would house the anchorages, and the varying demands of building one pier virtually on shore and another in the open sea. They studied the problems of supplying men and materials from both sides of the Gate, shipping steel, wire, and equipment in from the East, and hiring slow-to-learn novice workers.

They established the beginning of the cable operation as a control point and had everything else pivot around it. They wanted the same crews and equipment building both towers, and they wanted the unrelated construction of the foundations and the towers to dovetail, arriving at completion the same instant Roebling was ready to spin the cables. If they could arrange all that, construction of the roadway and the deck would follow quickly and economically.

There was always a danger with wind like the kind blowing in through the Golden Gate. The guys who had been around—they knew where to be, which sections had the lowest risk. They were the ones who lived long. The young, brave ones—it didn't matter. They'd work anywhere, proud of what they could do. They were the ones we lost. (George Albin)

By year's end they announced the schedule. The deadlines ranged from September 1, 1933 for the finishing of the south pier, to December 1, 1936 for completing the paving of the roadway. Construction was to commence in January of 1933 with the on-shore foundations, specifically the pylons and anchorages, the work contracted to Barrett & Hilp.

The pylons—four hollow reinforced concrete shafts, two on each side—would support the roadway at the bridgeheads and guide the cables to their moorings. Initially the pylons would be built to three-quarters their final height of 250 feet, so as not to interfere with the cable spinning still two years in the future. It was simple dry land construction.

Preparing the anchorages was more complex. These massive units were concrete buildings that would knit with the earth on either side of the Gate and hold secure the ends of the main cables. For that the anchorages had to be virtually buried in the ground.

Excavation for the anchorages began in January with power shovels and blasting powder alternately cutting into the flanks of the gorge. Pits deep enough to swallow a 12-story building were carved, two into the sandstone and shale hillside at Lime Point, and two into the serpentine slopes behind Fort Scott. Each of the four anchorages was then built in three interlocking pieces. Base blocks went down first, rectangular structures 60 feet wide, 170 feet long and 97 feet high. On top of them were the anchor blocks, the units that would ultimately hold the tiedown fixtures for the cables. And on top of them were weight blocks. When finished, the anchorages were triply reinforced masses of concrete each built to resist 63,000,000 pounds of pull without budging so much as an inch.

The Marin pits were dug without delay to a depth of 78 feet on the ocean side and 84.5 on the Bay side. They were ready for pouring by late spring. The San Francisco operation lagged somewhat behind, slowed by the serpentine rock's resistance and the presence of old Fort Scott.

Rather than test the District's right of eminent domain, and because he admired the structure, Strauss insisted on preserving the fort, the only West Coast example of pre-Civil War military architecture. The fort was clearly in the way, so Strauss ordered a detour. He had the contractors widen and pave the area around the building. That meant dismantling hundreds of tons of granite rocks from the seawall, storing them until after the road was completed, and rebuilding the wall exactly as it had been before.

Even with the delay, the anchorages and pylons rose uneventfully through the first few months of 1933. The job was big but routine, carpenters building forms to hold the concrete so it would not belly out while curing, crews settling down reinforcing iron, men pouring and leveling the concrete.

It was hard work—low-paying, sometimes sporadic, and always insecure. Barrett & Hilp was not a particularly progressive firm. "We wore our union buttons under our hats," recalls one worker. And the company was concerned primarily with the economic strictures of a cost-plus contract.

Barrett & Hilp was a tough outfit to work for, no question about it. They had a deadline to make, and that was that. They had a saying in those days, "Eight for eight or out the gate," which meant if you couldn't put out eight hours of work for eight hours of pay, you were gone, didn't matter who you were.

That job separated the sheep from the goats. There was only two colored working for Barrett & Hilp, and I was one of them. There was a lot of workers who didn't want to work with a "nigger." Nobody ever said that to my face, mind you, because they knew I'd pound a two by four across the side of their head. But the attitude was there. The company was fair about it, though. Barrett & Hilp knew a man's color didn't have nothing to do with his ability. They laid out the work for you and said, "this is your project." Twenty minutes later, if you were still standing around scratching your head—well, you might get away with it for an hour, but after that you'd be gone.

There was a fence not far from where we worked. On the other side were all the people waiting for somebody to get hurt or screw up so they could have a shot at the job. When the foreman came out of his shack and headed toward the fence, you'd think they were going to mob the gate. He would walk toward them, crook his finger at some guy, and say, "You!" and that was how people got hired. Those new people didn't walk to the job, they ran. And when they got started they worked. If one crew could put out twenty-five forms a day, the other crews had better be able to put out as many. If not, the foreman'd hand the laggers their time cards and head for the fence. (William D. Smith)

At approximately the same time Barrett & Hilp was raising the anchorages and pylons, Pacific Bridge was beginning construction of the tower piers, that part of the foundation in the ocean itself. With this step, the building of the Golden Gate Bridge became a pioneer venture.

Never before had piers been sunk in the open sea. The problems posed by the San Francisco side alone were without precedent in civil engineering. The south pier had to be placed more than 1,100 feet offshore, amid the full fury of marauding tides, at a bends-inviting depth ranging to 100 feet. It had to be built without a breakwater because a breakwater would have cost more than the pier itself.

The Marin pier, by contrast, was relatively uncomplicated. Soundings taken by the E. J. Longyear Company in 1929 revealed a subaqueous ledge some 20 feet below the waterline at the rocky lip of Lime Point. It was a spacious underwater mesa stout enough to support the full weight of the north tower.

Carpenter crews built a vast network of wooden forms in the anchorage pits. On call only for formwork, they were laid off when concrete was being poured. "A two, two-and-a-half day work week was pretty good for us in those days," recalls one Barret & Hilp worker.

Construction began with the excavation of the cavernous pits on each side of the channel for housing the mammoth concrete anchorages. The San Francisco pit, pictured here, was dug just behind Fort Winfield Scott.

By summer of 1933, the pylons (foreground) were nearly 187 feet tall, the height they would remain until after the cables were spun. The anchorages (background) were also close to their cable-spinning dimensions, approximately two-thirds their eventual height.

The first section of the anchorage was the base block, a thick layer of concrete with a stepped-off surface that knitted to the bottom of the pit. At the far end of the base block, workers installed steel anchor chain girders.

The anchor chains served as support members for eyebar chains which would eventually hold the cables. Each main cable would be composed of 61 bundles, or "strands," of bridge wire each wrapped around a special fixture which attached to an eyebar. The 61 eyebars at each anchorage were placed at angles pointing slightly upward.

With the eyebars in place, Barrett & Hilp carpenters moved in and built forms for pouring the next level in the anchorage, the anchor block.

The anchor block buried the eyebars up to their necks, 130 feet deep in solid concrete. At this stage, each anchorage weighed approximately 50,000 tons and measured 60 feet wide by 170 long and 97 high.

An aerial view of the Marin foundation reveals the anchorage and pylons ready to receive the cables. The anchorage eyebars are fully enclosed by cured concrete.

This view from Marin shows the anchorages, pylons, and tower in the summer of 1935, just before J. A. Roebling & Sons began spinning cable for the span. The equipment on the towertops and anchorage housing was used to string catwalk foundation ropes across the channel. The batching plant in the foreground was constructed specifically for foundation work on the Marin side of the Bridge. There was a duplicate operation on the San Francisco shore.

The towers were erected by June of 1935, and footbridges were installed by September. For the next nine months, the main cables were spun across the Golden Gate. The pylons guided the cable ends along their final approach to the eyebar fixtures. As much as 1,000 miles of bridge wire raced back and forth across the span through portals in the pylons in an eight-hour shift.

By midsummer of 1935, the once unbridgeable gorge was visibly conquered. Nearly 389,000 cubic yards of concrete and 44,000 tons of structural steel had been converted into anchorages, pylons, piers, and towers, the dimensions of which were unprecedented in the history of civil engineering.

Cable-spinning began and ended at the face of the anchor blocks. There the 1,600-pound coils of bridge wire were fed by unwinding bobbins to the spinning wheels. And there the individual strands of spun wire were rooted to the "strand shoes," the anchoring fixtures coupled to the eyebars.

It was also close enough to shore that the building site could be completely enclosed by a dewatered fortress.

For that job, the engineers ordered the skeleton of a three-sided cofferdam built at the Sausalito shipyard of Madden & Lewis. Upon completion the hollow frame cribbing, knuckled at the joints like the adjoining walls of a log cabin, was towed to the pier site by barge, lashed together with beams, filled with broken rock, and sunk to the bottom. It was connected to the natural wall of the Marin hillside with a rock dyke, then its three remaining sides were covered with interlocking panels of sheet steel. At first the dam was calked by jamming the seams with sacks of wheat which would swell when soaked by inrushing water. Later more accessible red dirt from nearby hills was used instead.

We put that cofferdam together on barges over at Madden and Lewis. We built it upside down, so that when they took it out to the site they could just turn it over in the water, and it'd be right in place. Only thing is, they hadn't figured on how deep it would sink once it hit water. When they towed it out there and upturned it, it sunk so deep in the mud they couldn't drag it out.

They had to call in a couple of tugs from San Francisco to pull that big thing out of the mud. But you know, when they eventually turned it over and filled it with rocking, it fit almost perfectly. The engineers had that dam figured so carefully that it was almost level when it hit the bottom. The floor of the Bay where they set it was 32 feet in one corner and 20 feet on the opposite side. It was tapered, and we had built the dam to exactly the same degree of taper. (Francis Baptiste)

Once the dam was in place the area inside was pumped dry, exposing a huge dank quarry with hundreds of fish—cod, bass, crab and small abalone—squiggling frantically across the muddy bottom. Workers took time out briefly to fill gunny sacks with fish, then they began preparations for the pier. For the next several weeks they blasted with dynamite and dug with jackhammers, excavating 15 feet into solid rock.

The pier was to measure 80 by 160 feet, rising 65 feet from bedrock, 44 of that above the water line. When the workers had cleared an area inside the dam half again the size of a football field, the carpenters entered and began building forms. After them followed rebar crews installing dense forests of reinforcing iron, wire mesh set flat against the ground, and narrow posts of ribbed steel. Then fleets of body-mixer trucks, a novelty for the times, rolled toward the site from batching plants built beside the channel, mixing concrete as they moved.

They really poured, yessir. A regular river of cement could come down there. Those trucks would dump into a big chute that pointed in the forms, and the two guys working the chute really had a job. They had to be careful that the bottom of the chute didn't hit right where they were pouring. If they had a mound there, and they hit it with new cement, the cement would back up over the top like water from a busted main. I got buried that way once. Oh Jesus, yeah.

I was down at the bottom getting ready for a pour—on my hands and knees cleaning up all the debris that the carpenters had dropped on that reinforcing iron, what we called the rattrap. I had a five gallon bucket and was throwing in all these pieces of wood and crap. Right about then, somebody wasn't paying attention and the chute flooded with cement. It backed up over the top and spilled out from about 20 feet in the air right on top of me. They used that huge 2½ rock in that cement, and everywhere the rocks hit me I got cut. I wasn't knocked out because I had a hard hat on, but everywhere I was cut, I got cement goo under my skin. My back was one big water blister. (Frenchy Gales)

The plan for pouring involved separating the huge form into quarters and filling one quarter to a level of four feet before crossing diagonally to the next, and so on until all four rooms were poured.

All through the spring of 1933, the north pier climbed, four foot layer after four foot layer, toward its final height of 65 feet. As it rose, seventy-eight massive steel dowels were planted 56 feet deep inside the hardening pier. These would stick up above the top of the structure and become the uniting bond between the pier and the tower, 107 tons of anchoring steel rods ultimately supporting the main material of the superstructure.

The dowels were secure by the middle of June, and by June 29 the north pier was officially finished. Pacific Bridge handed it over to the District sixty-four days ahead of schedule. It had been so smooth an operation that no one foresaw the problems that were lurking just 4,200 feet away on the other side of the narrows.

All the early opposition to the Bridge seemed to center around the alleged instability of the bedrock north of Fort Point. Now, years later, though construction of the south pier was underway, the site was still causing difficulties. In the fall of 1929 it had taken the Longyear Company three months just to sound the channel. Drilling crews wore through diamond bits as though they were knife points trying to cut through a grindstone. The exploration barge fared little better, rolling constantly from the currents (which often reached 8 miles an hour) and rising or falling as much as 7 feet during the changing of the tides. "One of the fellows told me many times of being out on that barge," remembers one of the Pacific Bridge workers. "He said sometimes it moved up and down so rough, he had to hang on to the boat just to keep from falling off the world."

The Longyear barge was moored with four 9-ton and two 6-ton anchors, but still there were occasions when waves pitched the boat so violently that work had to be suspended for days at a time. It was as though Nature itself were opposed to the construction of the concrete pier.

Ever since 1869, when the Brooklyn Bridge was built across New York's East River, the standard method for pouring an underwater foundation involved the use of a floating caisson. This immense cellular structure, in effect a multi-story metal building, had a circumference matching that of the proposed pier. It would be assembled in dry dock, floated to the site, then anchored into place. Once situated, the caisson would be used as both a pier base and an excavation bell. The top surface would be latticed with forms and layered with concrete until the caisson itself sunk flush to the bottom. The chambers inside, filled with compressed air, would act as giant workrooms in which workers could move around as they excavated bedrock, smoothed foundation floors, and disposed of broken rock and waste. When the workers finished, the caisson would be sealed with concrete, becoming the reinforcing iron for the footing of the bridge.

Strauss, Paine, and the Pacific Bridge engineers employed the caisson theory as a matter of accepted precedent. For waging war with the elements of the Golden Gate, however, they had to customize it carefully. Somebody—either Strauss or Paine—happened upon the idea of building a massive elliptical fender around the work site, a sturdy concrete wall sunk in benching carved expressly for that purpose. It would be raised to 15 feet above sea level and used to shield the construction zone from the force of the tides. Choppy channel waters would also preclude the use of construction barges; for more stable access to the workstage, they devised a trestle, a working wharf reaching 1,125 feet from the shore at Fort Point.

Building the trestle was simple enough. As the first step in the south pier project, it was begun early in the spring of 1933. Explosive charges detonated through hollow pipes dug postholes in the adamant rock along both sides of the trestle path. Into these holes were inserted the 16-inch I-beams that became support girders for the timberwork, which, when completed, would house electrical lines, telephone wires, a railroad track, a six-inch water pipe, two compressed air lines, and ample room for a steady flow of truck traffic.

At the same time the wharf was being raised, the channel was being excavated. Originally, the blueprint called for all digging to be done from inside the caisson. Pacific Bridge soon scrapped that plan. The pier site was actually a sloping underwater ledge, ranging in depth from 60 to 80 feet. After assaying Longyear's core samples, geologists Lawson and Sedgwick recommended taking the floor down to 100 feet. But digging that far into rock from inside a caisson would be too cumbersome and costly a task. Instead, Pacific Bridge received permission to blast down to the recommended depth.

From the derrick barge *Ajax* they dropped cylindrical bombs through hollow pipes suspended from a timber-frame tripod. Because of the hardness of the rock, the bombing had to be conducted in two stages. First, small explosives charged with three pounds of dynamite were dropped into holes started by drill bit. It would take an average of fifteen of these bombs in each hole to blast out the pilot apertures. Then, large bombs 20 feet long and filled with 200 pounds of dynamite were lowered into the pilot holes. When six of these large bombs had been tamped into as many holes, marine divers would take wire leads from the detonaters located on board the barge and connect them to the bombheads. The *Ajax* would wait for the divers to return, then move out of range and set off the charge.

When the entire surface was one blasting level (15 feet) deep in shattered rock, the *Ajax* would send down a five-cubic-yard clamshell bucket to scoop up the rubble and deposit it on a refuse barge.

The pier site was moving steadily downward and the trestle was nearing completion when San Francisco's summer fog began to hang dour over construction—a wet, gray harbinger of the trouble that would mount from the seas for the next six months.

On the night of August 14, a McCormick Line freighter, the *Sydney M. Hauptmann*, wandered 2,000 feet off course, ignored the bleats of panicking foghorns, and hit the trestle at its mid-sections, snapping it in two. Damage to the freighter was minimal, but $10,000 worth of bridge construction was destroyed in the accident. More importantly, the mishap tacked several months onto a work schedule that had allowed for no such delay.

On October 31, an unexpected storm raged in off the Pacific, and in its wrath destroyed part of the rebuilt trestle and the entire first section of the fender. On December 14, another, more violent storm hammered at the same point, this time twisting 800 feet of refashioned trestle and fender into ueless scraps. Five months were lost and the trestle had to be rebuilt at a cost of $100,000.

Finally, in the spring of 1934, the slow process of building the fender began anew. In light of the weather, the original plan was modified to make the fender connect with the base of the pier. Instead of building it in 60 feet of water, the engineers decided to bring it down to 100 feet also. And instead of building it in vertical sections, one at a time from bottom to top, they planned to construct it in horizontal layers. Sections 30 feet square would be set all around the bottom, with succeeding layers stacked on top, much the same as one would erect a brick wall.

A good portion of the initial work had to be done by Pacific Bridge divers, brought down from the Columbia River. They were specialists who could work fast and sure in swirling waters so murky they could not even see their hands while they toiled.

The Golden Gate—I don't know when they started it, but when they called me down, they already had their piling drove and the trestle work going out. That was in the spring of '34. When I got down there, they were just beginning to lower the forms for that fender ring.

It was treacherous as hell diving for that job. You could only work at slack water, high water slack or low water slack. The tide would see to that. Even when it was slack, it could be tough. While it was slowing down or speeding up, it could still move fast enough to come awhipping in and knock you for a loop.

I think we only had an hour and fifteen minutes we could work down there at any one time. And then we would have to come up so fast we couldn't recompress in the water. The safest way to recompress is to do it naturally, on the way up. But what're you gonna do if the current is coming in so fast it'd knock you right out of the water?

They brought us up after a dive, and there wouldn't even be enough time for us to take off our suits. My brother-in-law, Ike Porter, he would tend for me most of the time. Soon as I got up, he'd yank my helmet off. We'd leave it right there on deck while Ike and me and all my riggings hopped on back of a pickup so's it could speed us to the recompression chamber down there at the end of the dock. A few times, I'd feel the bends coming on just before I made that chamber. But you get educated pretty fast on that subject. You know how long you've been down there. Sometimes you just have to make a run for it or you're a dead pigeon. That's one thing you don't crowd. That gets you too close to the old man. (Bob Patching)

The divers first had to smooth out the jagged corners of the excavated pier site, so the fender could be set onto an even surface. Using hoses capable of exerting 500 pounds of hydraulic pressure, they worked blind, assessing by feel as they smoothed the floor. Next, they had to help align the huge steel forms that would shape the wet cement.

For working on that fender ring, we dove off the trestle. They had one of those whirligigs attached to a tripod scaffold. We'd hook onto that whirligig, and the operator would hoist us up and move us out over where we were supposed to work. We'd be about 20 feet above water, then he'd let us go.

Our job was to line up sheet steel panels with the I-beams that were sunk in the bottom as guides.

They were for framing in the forms. We'd go down and take a hook on a line from up top and put it around the I-beam. Then they'd lower a panel, and we'd bolt it to the I-beam with one-inch bolts so's it couldn't slip off.

Well, that went along pretty good, except once these forms were in, they didn't always fit the bottom all the way around. The bosses didn't want concrete running out underneath the fender forms. If the cement escaped, all you'd have inside that part of the base of the pier would be sand and gravel. So, we got everything ready for pouring except for some holes between the bottom and the forms. The bosses got the idea of putting some canvas in them holes—take some nails and nail big sheets of canvas down to plug them up. Well, we could see right away that was no soap. If we'd a brung canvas down there, the first thing you know here would come some swell and wrap that canvas around a diver like he was in a sack and take him out to sea. The next idea was a good one. They went and got some chicken wire and had us bring it down to patch those forms with. Yessir, we used chicken wire. That's what the very bottom of the Bridge is held together with. (Bob Patching)

The fender was to be constructed in twenty-two sections that would eventually circle the pier. For insertion of the caisson, however, the eight sections facing due east had to be left open. By the fall of 1934, the fender was complete in its horseshoe configuration, ready for the caisson to be towed in from Moore dry dock in Alameda.

On the eighth of October, an Indian summer afternoon, the caisson was floated toward Fort Point. The water was millpond smooth when they drifted the big building into place, but during the night Nature let loose one of its inexplicable Pacific storms, rolling in swells so violent that the caisson was pitching and rocking uncontrollably when the men reported to work the next day. With every swell it wrenched looser from the blocks and tackles straining to hold it in place.

Divers were sent down to insert steel anchoring posts, but their efforts failed to constrain the raging structure for long. The storm blew all day and into the night. By midnight the weather showed no sign of improving, and Joseph Strauss was called from his bed to come to Fort Point and meet with Pacific Bridge engineers.

Clifford Paine was back east on business, so Strauss had to bear the burden alone. In response he made one of civil engineering's more innovative decisions. It would take two weeks to close up the fender, he reasoned, by which time little would be left of either the caisson or the ring. This forced him to rethink the need for a caisson. It could be removed, the fender enclosed and joined to a thick slab of foundation cement. Then the entire area would be dewatered, and the footing built much the same way Pacific Bridge had constructed the trouble-free north pier.

Pacific Bridge engineers seconded Strauss's thinking, and the decision was made. It remained only for the caisson to be released and led safely out.

Chris Hansen, Johnny Bacon, and me were all standing around, looking at that caisson bobbing up and down, and knowing all the time what was coming up. First thing somebody had to do was pull out those two big posts that was supposed to be holding the thing in place. They was connected to half-inch bolts that was splayed in half and bent sideways—they was actually two big cotter pins.

I forget how it came about, but Chris says, "Well, Bob, that's a dangerous one, getting them pins out." I says, "You ain't kidding me one damn bit. That water is a falling fast enough to knock that thing to pieces. It's a going like a bat out of hell, and you got to get right in the middle of it to pull the pins." He says, "Well, I'll try for one, if you'll get the other." I says, "Which do you want?" And that's how we got to going down there.

Well, we rigged ourselves with extra lines. We tied ourselves back in such a way that we could loosen the lines, so's that if we got pulled up against something and couldn't get away, we could get help from upstairs. We was working in a vacuum down there. We were down damn near 50 feet, and every time you go down 29 feet you double your atmospheric pressure. Well, that's strong enough it can hold you smack against a wall, and you can't move. But you got this consolation: it can hold you only so long as the swell is against you; when it comes back, you ain't a going to stay there unless you got something to hang on to.

I had figured it pretty close; it was just like I thought down there. Of all the experience I had as a diver, I used the bulk of it right there on that dive.

I got down there, strapped myself in good and tight, reached over and got hold of my pin and commenced to bend it back and forth. You can't see down there, especially when the swells are a coming in, so I just had to feel my way. Well, both of us worked on those pins long enough to finally snap them, then we got the hell back upstairs so's they could pull the posts and float that caisson away. (Bob Patching)

The story of the ill-fated caisson did not conclude with the divers releasing it for easy removal. The tugs from Oakland retrieved it all right, but, like a poorly trained dog, it would cooperate only to a point. En route to dry dock the huge steel structure broke its two lines and careened into a pier of the as yet unfinished Bay Bridge. After recovery from that escapade, the caisson sat in the Moore yard for nearly four months while Pacific Bridge decided what to do with it. The custom-built structure was useless for anything other than its original purpose, and eventually Pacific Bridge decided to tow it far out to sea, stuff it full of dynamite and blow it up.

On March 4, the caisson was put under the care of the tugs *Kodiak* and *Rustler*. They pulled it out through the Gate and towed it twelve hours, making only five miles in all that time. Shortly after nightfall a savage wave hit the convoy, and the explosives-laden caisson lurched backward and snapped the tow ropes. Most of the night it drifted malevolently back toward the shipping lanes, a bloated sea mine gone mad. But by dawn the tugs had managed to resecure it without incident. They towed it thirty more miles before letting it go and detonating the charges. The entire saga was summed up by an unlettered seaman. "Damn box," said Tony, the tugboat captain, upon return to shore, "she go boom and sink."

About a month later a new bridge opponent surfaced, this one human—a man whose allegations sent the District back to political warfare. Seventy-eight-year-old Bailey Willis, a white-maned professor emeritus of geology from Stanford, stormed the periphery of the bridge project, ranting that the south pier bedrock was too weak to support the prodigious weight of the tower.

"It would appear that there is reasonable ground to fear a structure thus supported must sooner or later be destroyed by landslide," he argued, "either as a result of prolonged stress and fatigue of material, or as the result of earthquake, shock, or both."

Willis' alarmist opinions were first seen in an unsolicited report he prepared in April of 1934. Armed with figures and conclusions, he secured an interview with District President Filmer, who invited along General Manager Reed and Building Committee Chairman Keesling. They listened to Willis then consigned his charges to further study. By late summer, Willis had expanded his harangue, writing letters to Chief Engineer Strauss and Consulting Engineer Derleth. The matter was eventually passed on to geologist Andrew Lawson, who responded by denouncing Willis as a mountebank and charging the District $250 for the time spent in arriving at that opinion. For his part, Strauss refused to discuss engineering problems with an outsider, and the fight was on.

More letters from Willis, this time to participants in bridge finance, began to create discord. The directors decided to conduct an open hearing. Three sessions were needed (September 28, October 22, and October 30) for Willis to air his charges and propose as a solution excavating the pier site to 250 feet below its present depth of 100 feet. At $10,000 a foot, it would be frighteningly expensive.

Fortunately, Willis could not document his contentions. In wild flurries of research, he had ignored the official records, relying instead on some out-of-date geodetic charts, which he misread. His calculations placed the south pier 178 feet north and 120 feet west of where it actually was; he was 100 feet off as to the water's depth. "The record," said hearings Chairman Keesling while summing up the committee's findings, "discloses inaccuracy where accuracy is demanded of a man of science. . . . Professor Willis had not substantiated his conclusions."

And so in mid-January, 1935, the final episode of resistance to the south pier ended. The Willis matter was concluded just one week after the completed fender ring was dewatered, and approximately six weeks before the pier itself reached its final height of 44 feet above sea level. After so many years, political opposition was finally silenced. And after exerting so much of its wrath, Nature was finally allowing the south pier to be built. With the successful completion of that footing in January of 1935, the most difficult obstacle to the project had been surmounted.■

The Marin tower pier was to be built near the shore, where the channel was no deeper than 32 feet. A cofferdam was constructed, sunk flush to the bottom, then dewatered so the pier could be built in the dry.

Left
On that tower pier, we'd pour anywhere from 900 to 1,200 yards a shift. They'd get the cement going so fast it'd pile up, and that's where we'd come in —guys with rubber boots jumping on it and moving it around to make sure it'd flow all even. (Frenchy Gales)

Right
The Marin pier was finished on June 29, 1933, 64 days ahead of schedule. Solid concrete, it measured 80 by 160 feet and rose 65 feet from bedrock, 44 of that above the water. Angle irons were buried 56 feet deep in the pier to connect it to the tower.

The San Francisco pier was to be built 1,100 feet offshore. For access to the site, a working wharf was constructed and an elliptical, protective fender was built around the pier area. Here, a construction barge is lowering one of the initial fender forms into position.

With the inbound tides, foaming waves surge and eddy around Fort Point, slapping against the hills and breaking violently onto the Presidio beaches. The San Francisco pier, set on a bare rock underwater ledge, would be fully exposed to heavy seas and a treacherous surf.

Lost in the fog in mid-August, 1933, an outbound freighter plowed into the midsections of the newly erected pier, causing substantial damage. In late October, a Pacific storm demolished 800 feet of the rebuilt structure. Five months were lost as the wharf was finally constructed to withstand any challenge.

The pounding waters of the channel forced Strauss and Paine to abandon the use of derrick barges and build the construction wharf instead. With every inbound tide, the wisdom of this departure from standard practice was confirmed.

Left
Divers did all the work on the bottom for the San Francisco pier. They wore what they called hardhat diving dress—a rubber and canvas suit with a rubber collar, a studded breastplate, and a helmet that clamped onto the collar and was twisted a quarter turn before a dive tender locked it into position.
(Roy Brownell)

Right
To reach stable rock the south pier site had to be excavated as deep as 40 feet below the channel bottom. Most of the digging was done by cylindrical bombs filled with 200 pounds of blasting powder. Here, a Pacific Bridge worker, balanced on explosives crates, is loading the bombshells.

Right
Underwater, marine divers connected wire leads from detonators on the barge to the bomb heads then surfaced. When the divers were back on board, the bombs were triggered. One blasting operation could excavate to a depth of 15 feet.

Left
Pilot holes were blasted in the channel floor and large cylinder bombs, pictured here, were lowered through hollow pipes on a derrick barge. When six of the big bombs had been stuffed into each of the holes, the blasting was set to begin.

Right
The big bombs could reduce the channel floor to rubble, but they rippled the surface of the Bay no more violently than the passing steamships.

Left
Originally, the south pier plans called for use of the giant steel caisson pictured here. Built in dry dock, it was floated through the open east end of the horseshoe-shaped fender. There it was to become both a building platform and the reinforcing iron for the south pier.

Right
After a storm nearly destroyed both caisson and fender, the caisson was removed and the fender enclosed. Fender and bedrock were then joined with concrete, and the area inside was dewatered so construction could proceed in the dry.

Left
With the pier site dewatered, carpenters moved in and built forms for pouring concrete. The smokestack-like pipes are the hollow vertical shafts of inspection bells, the steel chambers used by engineers and geologists to monitor the condition of bedrock.

Right
Foundation concrete, nearly 65 feet thick, rose from bedrock to become the pier base. All the pouring was accomplished through tremies—steel trunks fitted with funnel-top hoppers—that men controlled with blocks-and-tackles.

Prior to their installation, the inspection bells were warehoused near Fort Point. They were equipped with air locks for maintaining positive air pressure—even at depths some 100 feet below the surface of the Bay.

Before the pier itself was poured, the foundation rock was inspected first hand. When geologists confirmed that the bedrock could indeed support the weight of the span, the eight inspection chambers and shafts were filled with concrete, and work on the pier was begun.

An aerial view of the rising Bridge. Fort Point, once the lonely sentinal of the Gate, is obscured by the mammoth south end foundations, the anchorages, the pylons, the elliptical fender and the pier in progress.

All obstacles cleared, the south pier proceeded apace. Each position of the tremie could feed an area of 900 square feet, and concreting at a rate of 100 cubic yards an hour was common.

On January 3, 1935, the SF pier reached its final height of 44 feet above the water. Within days, the site was teeming with new activity—Pacific Bridge dismantling equipment, and Bethlehem bringing in men and materials for construction of the south tower.

Towers: The Steel Is Raised

Give us the plans, and we will build a bridge to heaven or to hell. —A steelworker motto

A Majestic Doorway

By 1930 the engineers were unanimous in their conviction that the beauty of the Golden Gate precluded all but the most aesthetic bridge design. They infused the project so thoroughly with this sentiment that the entire staff became intent on finding the shapes and forms that would surpass the drab geometry of function. Chief Assistant Engineer Clifford Paine and staff architect Irving Morrow envisioned a structure whose profile would excite the senses gradually to full admiration, the same way the arc of a well-thrown javelin slowly stimulates its viewers just before they applaud the grace of its flight.

The most dramatic element of the Golden Gate Bridge would be its singular height, particularly when viewed in the context of its surroundings. Cables 4,200 feet long in the center span would form a curve 475 feet deep, requiring towers that would rise 690 feet from the tops of the piers, 746 feet from the water line. This would make them 232 feet higher than the towers of the San Francisco-Oakland Bay Bridge, one third as high as Mount Tamalpais, the peak of Marin's coastal highlands, and twice as high as the Russ Building, then the tallest structure in San Francisco's skyline.

Paine and Morrow both believed the Bridge would appear most majestic if its height were emphasized subtlely; hence, they designed towers that slimmed as they climbed, giving the impression of an infinite ladder to the clouds. "Where great heights are involved," reasoned Paine with a dryness belying his creativity, "the common architectural practice of enlarging the scale of details as their distance from the ground increases tends to emphasize detail at the expense of the whole mass." The opposite principle would be employed on the towers of the Golden Gate Bridge."

The towers would be composed of shafts, or legs, that would be joined by cross-bracing along the 200 or so feet from the pier to the roadway and, above that, by horizontal braces, called portal struts, spaced along the 500 feet from the roadway to the tower tops. The shafts would be built not as solid columns of steel, but as clusters of hollow cells. Fashioned from inch-thick plates, these cells would measure 42 inches square and 35 feet high. Each shaft would be formed at its base by a honeycomb of 97 cells. As the shafts rose, the number of perimeter cells would drop off at intervals corresponding to the placement of the braces, until only twenty-one cells remained to make up the tops of the towers.

The narrowing effect would be enhanced visually by the four portal openings, which would likewise decrease in dimension as they traveled upward. Portal bracing itself was an unusual choice. Paine saw no structural difference between it and the more conventional X-bracing; both he and Morrow found it pleasing to the eye. So did Strauss, who had retained for himself final assent on all matters of design. "This bridge," he said exultantly, "is perhaps the first in which the importance of the new motif of stepped-off towers has been recognized and applied. The bridge is also the first in which the network of transverse braces between the tower posts is eliminated, and the towers portal-braced throughout, making the tower effect that of a majestic doorway."

In another way, the towers were shaped by the need for a combination of flexibility and strength. Each tower would have to support itself, part of the cables, and part of the roadway (about 75,000,000 pounds of dead weight) plus as much as 9,500,000 pounds of live weight should the bridge ever be lined end-to-end with loaded trucks. They would also need the elasticity to bend with the expansion and contraction of the cables, and to yield in the face of prevailing winds.

Shortly after the preliminary design was submitted in February of 1930, responsibility for stress analysis was assigned to two men: Frederick Lienhard, from the office of Leon Moisseiff (himself among the world's foremost authorities on the subject), and Charles Ellis, then a vice-president in the Strauss firm. Lienhard and Ellis set up a testing laboratory at Princeton University in New Jersey. They authorized the Budd Manufacturing Company of Philadelphia to build a model tower of stainless steel to 1/56 scale. Then they proceeded to simulate and compute all the pressures the actual towers would face.

They determined the specifications for every piece of metal, including rivets. They devised standards for strength. They computed the flexibility that would allow the towers to bend 18 inches toward the channel with weather-induced contraction of the cables, and 22 inches toward the shore with expansion. They calculated wind stresses and prescribed towers that could move up to 13 inches in lateral directions. They concluded their analysis with a set of thirty-three simultaneous equations involving up to a half dozen unknown quantities each. In effect, they came up with algebraic language that was translated by the engineering staff into a final design and, eventually into orders for sizes, shapes, and quantities of steel.

All the steel for the towers was prefabricated, forged, and tested in Bethlehem's foundries at Pottstown and Steelton, Pennsylvania. Beginning in March of 1933, it was brought by flatcar to Philadelphia, then transferred to barges which shipped it down the Atlantic Coast, around the Gulf of Mexico, through the Panama Canal, up the western shore of Mexico and California, through the Golden Gate, and on to Alameda. There it was stored along with derricks and rigs brought west from the George Washington Bridge until the Marin pier was completed in the summer of 1933.

Bethlehem's involvement in the Bridge construction brought ironworkers to the Golden Gate for the first time. These were nomadic men who would travel gladly from New York to Detroit to Panama to Saudia Arabia if that were the path leading to the big jobs in structural steel. They moved together in small cliques, bonded fast by their familiarity and interdependence. They were joined in spirit by their pride in functioning without fear under conditions that would have most men quivering with inhibition.

Bridgemen, as they liked to be called, were a unique species in American labor. Joseph Strauss once described them as George Patton might have talked about select infantry in his 3rd Army: "There is an ancient saying among bridgemen," he intoned, his voice filled with awe, "that the bridge demands its life. Among modern builders, the rule of thumb is one life for every million dollars worth of construction. That seems an appalling sacrifice, and one would think it next to impossible to find men to take the risk. But the very reverse is true. Bridgemen are a breed unto themselves, strange migratory birds with an uncanny ability to sense the next big job. There are always more men seeking work than there are jobs, just as there are always more men than are needed to fight a war."

Ironworkers of the day were generally excited by the acrobatics of high steel labor. The risk of death

Left
Steel for the Bridge was fabricated back east and shipped to Bethlehem's warehouse and assembly plant in Alameda. There, the pieces were catalogued and stacked, ready to be barged to the construction site in loads of 500 tons.

Right
The towers were designed for both flexibility and strength and thus were built from hollow steel cells honeycombed together. Each cell was identical to the one pictured here; they were made from inch-thick steel and measured 35 feet in height.

One of the few decorative touches in the span's design: fluted brackets were placed inside the upper corners of the tower portals. Only a small section of a tower, a single bracket is still imposing enough to dwarf a standard-sized flatcar.

was real, but for the most part dismissed as the price they paid for working adventurous jobs all over the world at premium salaries.

In the old days, nobody thought much about safety or saving lives, just beating that one man per million score. Nowadays, it's different. You don't have your Pinky Brinkleys or Wino Smittys, or Hook-nose Smittys, or Snot-nose Smittys, or Dirty-neck McCaffertys, or Bad-eye Gerlicks. Those were characters who wouldn't even think of using a ladder or a safety line while they skinned up a ten-foot column that was hundreds of feet in the air. They'd just climb up the sides of the steel by stepping on rows of rivets like they was rungs on a ladder, swing a leg over a girder, hang there, and make their cuts or put in their bolts. (Jay Hollcraft)

Ironwork had experienced a swashbuckling infancy and adolescence, but by the mid-1930's it was beginning to mature. The era of heedless daring was coming to an end, hastened in part by practices instituted on the Golden Gate. This bridge, the workers were soon to discover, would be one of the first enforcing rigid safety codes. As one of his minor though fervent ambitions, Strauss vowed to pare the "one life per million" formula to a record minimum. "On the Golden Gate Bridge," he said in a 1937 article for *The Saturday Evening Post*, "we had the idea that we could cheat death by providing every known safety device for workers [including one of the first requirements of hardhats and safety lines]. To the annoyance of the daredevils who loved to stunt at the end of the cables, far out in space, we fired any man we caught stunting on the job."

Up to that job on the Golden Gate, you were on your own. Nobody cared. If you were fool enough to clown around, you were your own fool. But if they caught you clowning out at the Golden Gate, you were fired. After that, the men themselves began to take a big interest in safety. Why the hell go out and commit suicide when the company was willing to take precautions to save your life?

You want to know how it used to be? One time I was working with a guy on one of the towers of the George Washington Bridge. It was lunch time, and we were sitting up there eating. There was a rig near us; it had one beam sticking out over the roadway. This guy asked me if I could walk the beam. I said I could, but I wasn't going to. He sat there and looked at it for a second, a 20-inch I-beam about 40 feet long, then got up and walked the whole way across it.

What the hell was I going to do, let him show me up? I got up and walked it too. It gets very tricky walking beams when there is no other steel to grab on to if you start to fall. It can get real lonesome. But I walked it just as fast and as sure as that other guy. The old captain, he was watching us from down on deck, and he got madder'n hell. He chewed us out good when he caught up to us. But that's all he did. We'd have tried that on the Golden Gate and nobody would have bothered to talk about it. We'd have been fired on the spot. (Harold McClain)

In preparation for the steelworkers, the cement surface of the Sausalito pier had been planed smooth, leveled with a grinding wheel so as not to vary more than 1/32 of an inch. Those variations were reduced even further, down to 3/1000 of an inch, applying thick coats of red lead paste. Then the base plates, steel slabs 5 inches thick, were set in. They were connected to the concrete with seventy-eight 4½-foot steel dowels, sunk in holes predrilled in both the pier and the steel. With the base plates secure, the transport ferry—a former rum running tug—began bringing ironworkers to the site. At the same time barges in Alameda started loading up huge pieces of steel for the tower now ready to rise.

On paper, tower raising is assembly-line smooth, but in fact nothing about the Golden Gate job allowed for complacent routine. The problems began with the weather; at best it was miserable, at worst it was treacherous.

You couldn't get used to the weather out there; working winters back East was easier on a guy than working summers out here. I started on the Gate Bridge in August. Right away, they gave me a job flashing signals to a crane operator, where I wasn't active physically. I nearly died from the cold. The second day out there, I found some tin and built barrier walls around—so just my head and shoulders stuck out—enough for the operator to see my signals. Later, one of the fellows brought me out an overcoat; it was so heavy I could hardly move my arms to give signals. One thing still sticks in my mind from that job—my feet were always cold.

The weather bothered those guys working the barges, too. One of my old pals had a job hooking on. Well, when those ocean swells come in, it can get pretty rough. The barge would anchor all right, but it would bob up and down, ten, twelve feet, sometimes more. The steel it was carrying weighed some 75 tons apiece, so it was trouble all the way. The guys hooking on would have a double hitch, and two big shackles and pins. They'd get that derrick as near right as they could, and just as it swung by, they'd shove those pins in, slam them home, and hunt for the boondocks. Because when that barge went down, it left steel dangling in the air; and when it came back up, it would hit the steel slam bang into the pile. My buddy quit the business after that job. His wife was putting too much pressure on him at home, and he decided he didn't want no more of steel work. (Harold McClain)

The men could not defeat the weather, but they did manage to co-exist, and the Sausalito tower made its slow ascent on schedule.

The steel was barged in from Alameda in loads of 500 tons, and transferred to a storage site on the pier. The interlocking units that averaged 65 tons apiece had been forged at Bethlehem's foundries and were predrilled with holes in buckshot patterns called riveting points. These holes were designed to correspond perfectly with holes in adjoining members, but they would align flush only when the steel was in position to receive all its anticipated burden. The engineers did not want the steel to slide easily into initial position if that meant it would buckle later under several thousand tons of pressure.

The towers were erected in three distinct phases. First, with the aid of an enormous two-headed derrick called a traveler, a "raising gang" signaled, guided, cajoled, and connected the pieces of steel. The traveler was a girderwork truss that fit snugly between the tower shafts. It was secured with moveable pins and beams, and was rigged with cables and pulleys that allowed it to crawl up the tower like some monstrous metal spider.

The traveler was outfitted with two separate booms, one servicing each shaft, both capable of lifting 85 tons. They would dangle their lines over a pile of steel, where "hooker-ons" would secure a piece with slings and pins then wave it off. Signal men employing a type of semaphore code would guide the dangling steel up slowly, helped often by men holding "tag" lines, who would stabilize the lift. Up on top, teams of "connectors," a man for each junction point, waited for the steel. When it approached they were supposed to guide it with one hand and wave directions with the other. Just as an end was eased into a joint, they would flash an "off" signal, then slam a temporary bolt into one of the holes and tighten it into place.

Sometimes, we'd be up there with a 65-ton piece of steel in the clear and ready to go down. I'd tell the boss to lower it, and the operator would literally drop it from the boom the last six or eight inches. And it might not go down. Sixty-five ton of steel, and it wouldn't budge an inch, it fit that tight. So we would take a 25-ton—or maybe it was 12½, I don't recall—airhammer and piledrive that sonofabitch into place. Still, it might be two inches off. Then, it was just a matter of putting more and more weight into them to force those girders down. Let's say you have a 65-ton piece and six pieces on top: six times 65 is 390, almost 400 ton of weight before those holes would align close enough so's you could drive a rivet in them.

I'll tell you how tight those things could fit. We set one of those pieces down one time. Kid Ford was working the derrick as engineer, and our boss was Eddie Wayne. We had big electric motors, and they could pull hell out of that steel. So, we dropped this thing down, and we couldn't get the top in. Turns out, we had put the top down first. Eddie Wayne says, "No problem, we'll just pull it out." But it wouldn't come out. Now, this big rig could lift near 100 ton. A 100-ton rig, and it couldn't lift out a 65-ton girder.

Finally, we had to jack it loose and when that thing came out, it jumped about 2 feet in the air, and set there scissoring until it came to rest about 14 inches higher than where we had set it. (Peanuts Coble)

Following the raising gang, crews of "bolterups" fitted the pieces together so steel was against steel with no possibility for slippage. They would fill about 40 to 50 per cent of the holes, all that would line up at that point, with drift pins—steel spikes tapered to a point—or with bolts.

When all the pieces were aligned, the "riveting gangs," each made up of a "heater," a "buckerup," two riveters, and maybe an apprentice or two arrived. They popped the bolts and pins and began filling the holes with rods of smoking hot steel.

The metal cooled to a permanent bond and was inspected rigorously. "The inspector, he took a small hammer and a washer," remembers one riveter. "He put the washer underneath the rivet and hit it on top. If the washer should jar or move, the rivet was loose. And when it was loose, you cut it out. Those guys were tough, those inspectors; they didn't miss much. And they inspected every rivet on that bridge."

The riveters' jobs often turned unwieldy because of the labrynthian design of the towers. The 3½-foot square shaft cells could accommodate only the two riveters, and even that was cramped. The buckerup worked in the next cell over; the heater and his forge were situated on scaffolds hanging outside the tower legs, often 100 to 120 feet away from his gang. He would send rivets to the buckerup through an aluminum hose, a pneumatic tube that snaked up, around, over, down, and around again, until it reached the destination cell. The buckerup would tong the hot rivet from a wire mesh trap, back it into a hole, and press against it with a dollybar. On the other side of the wall, the riveter would stand. As soon as he saw the glowing rivet shoved through the hole, he would jam it with his airjack and vibrate it carefully to a sculptured head.

It was a madhouse in those cells. You had a heater maybe 100 feet away, depending on where you happened to be, with that pneumatic tube coming toward you. You had to make sure the rivets were the right size—they changed with the plate thicknesses of the metal you were working on—and you had to make sure you were sticking it in the right hole. If you screwed up, it would only screw everybody up—the heater, because once you fire up a rivet you can use it only once, and the rest of us, because you would lose time and get behind. So, you're fighting all that, and you're fighting those coils of airhoses. It was pitch dark in those cells, and the ventilation was poor. You couldn't hear a thing except the noise of the riveting guns and the echoes. And there was many a time the light in your hard hat would go out. (George W. Albin)

Up on the scaffold, the heater would be stoking his forge, an open coal-fed fire nesting in a black iron tub. He would arrange his rivets in meticulous semi-circles, turning them over and over like so many broiling steaks, cooking them to the precise color of doneness. He would know in advance what sizes should be heated, and generally in what order. Beyond that, the only communication between heater and riveter was with their tools.

Lying right beside the heater, on the scaffold floor next to the rivet tube, was an airhose leading from the rivet gun to a compressor. As the heater cooked his rivets, he stood with his right foot on that hose. When the riveter was ready to begin a point, he would squeeze—brrt, brrrt—on his trigger. The heater would feel the vibration on the sole of his foot and start shooting rivets into the tube.

It was amazing how well that system worked. But there was this one situation. A heater didn't show up one day, and this gang needed another one fast. Now, this was hard times, believe me, and it was tough to get a job. And when somebody got one, they wanted to hold it. So this poor damn replacement heater, he gets situated on one of those big long struts between the tower legs, and he gets wrapped up in his work. As soon as he feels the guy rap for rivets, he starts sending them down. He's dipping into his keg, cooking them rivets, and shooting them down. He gets so wrapped up in what he's doing that he forgets to keep his foot on the hose. What the hell, he keeps shooting rivets and shooting rivets and shooting rivets. I guess he put about a half a keg down there.

All of a sudden, his riveter comes storming through the manhole, and this guy almost faints. He knew immediately what he'd done. I knew too, because I was part of the gang that had to come down and cut up his aluminum hose—this 120 foot length of tube was shot full of red hot rivets. This guy, when he saw the riveter, he looked at him, looked down at his forge that was still full of cooking rivets, reached down, picked up his lunch box and jacket, walked over to the elevator, and left. He didn't even wait for his wages. (Peanuts Coble)

By March of 1934, the ironworkers had built the north tower up to over 600 feet, and the first casualties of the project were beginning to report to the field hospital near the wharf at Fort Point. All the tower metal had been treated so liberally with red lead paint to retard corrosion that the rivet holes were covered by a viscous membrane of the stuff. And when the white hot rivets hit the membrane they touched off violet fumes of lead-tainted smoke that billowed poisonously through the unventilated cells. Workers inside the towers were beginning to turn up debilitated and worse.

There were sixty guys in the hospital at one time. And nobody knew what was causing it. The doctors were diagnosing it as appendicitis—sixty men all coming down with appendicitis at the same time. Finally, it came out that it was lead poisoning. It was awful, that stuff. Guys were losing their hair, their teeth; they were breathing shallow. There was guys who never went back to structural work after getting a dose of lead. (Whitey Pennala)

Treatment followed diagnosis, the victims filtered back to work or were replaced. To immunize against future epidemics, Strauss's and Bethlehem's staffs ordered some progressive changes. Compressed air was forced into the cells for ventilation; riveting gangs were asked to wear filtration masks (something many would pridefully refuse to do); bone phosphate pills were made readily available. Most importantly, from then on the holes in the steel yet to be riveted were reamed painstakingly by hand. Later, when steel for the San Francisco tower was being forged back East, it was treated with iron oxide instead of the red lead.

As it turned out, the threat from "leading" was probably the most serious setback in the construction of the north tower. In terms of time lost, it was not serious at all. By late spring of 1934, the tower was nearing full height. And when the final section of steel was riveted down in May, the tower was a 746-foot monument to iron work, 43,000,000 pounds of steel held together by 600,000 field-driven rivets, a structure every bit as strong and flexible as the engineers had hoped, a tower of boggling dimension built in iron mazes so intricate that men would get lost just trying to move from top to bottom.

About the time the Marin tower was finished the south pier was becoming bogged down in its delays. The slip in schedule proved unfortunate for the District, but gave Bethlehem an opportunity for further profit. In early summer of 1934, amid the setbacks with the San Francisco footing, the directors notified Bethlehem to halt production until November. Then it was to deliver all the steel for a completed structure nine months later. The original terms of the contract had anticipated that both towers would be up by January of 1935; that was the date around which Bethlehem had reserved its facilities and hired its men. Now the District was postponing that schedule for six months. Justifiably, the company balked, claiming it could not guarantee compliance with the new schedule. It would be willing to try, a spokesman said, if the District would pay a performance guarantee of $1,500 for every day under schedule that the tower was completed.

The directors were irate, until one of them, Hugo Newhouse, observed that the District would pay $4,600 in interest every day the project was set back. Under that logic, the rest saw Bethlehem's paid acceleration as a savings of $3,100 a day in interest. They ratified the agreement with a grumble.

When it came time to execute the terms, Bethlehem delivered. The north tower had been erected in just under ten months, taking from August of 1933 to May of 1934. With men and techniques refined in the crucible of that experience, Bethlehem finished the southern twin in just 101 working days on the 28th of June, 1935. There was only one incident of note in the building of the San Francisco tower. An earthquake rumbled through the San Andreas Fault in early June, setting the near-finished structure vibrating as violently as if it were some giant tuning fork that had just been struck by a mammoth and malicious rubber hammer.

I was up on the tower when it happened, and I guess there were twelve or fourteen others up there too. I was walking across—I didn't hear nothing, and all of a sudden, I felt like I was tipping off to one side. I sat down and shook my head. I thought I was dizzy or something. This friend of mine, he came along and asked what's the matter. I said, "Jesus, I feel like I'm swaying." He said, "It ain't you, the goddamn tower is swaying." I got up, and sure enough, this whole structure was swinging like a hammock.

Guys were going crazy. Some were climbing down through the cells. One guy grabbed some old gloves, put them on, and slid down a derrick cable, took a chance on killing himself—those derrick lines were all greased up, you know. Most of us stayed up there while she was going way over, first to one side, then the other. The elevator that ran up and down outside the tower shaft was about halfway up and was swaying away from the tower then coming back and banging against it. The poor guy inside was throwing up all over himself. Guys on top were throwing up too. After it stopped swaying, we all went down as fast as we could and got first aid for a sick stomach—I think it was a shot of whiskey.

The next day, we went back to work, and we got another one. Those quakes scared the bejesus out of me. I fell into the net on the backspan once, but that was nothing compared to being on the very top of that tower in an earthquake. And that next day, we was all up there talking about yesterday, and here comes another one. Shit. And you know, the engineers must have known there was going to be a second one, because they had guys from the University of California stationed down at the bottom to measure the sway of the tower. I talked to one guy later; he told me it swayed 30 feet. Thirty feet ain't much 746 feet up, but it's enough to scare the Jesus out of you if you're the one who's up that high. (Frenchy Gales)

It was fitting that Nature should have sent this one last shudder through the south end construction; though, compared to previous disasters, it was a fraternal, almost friendly nudge. It caused no damage, cost no loss of time; in fact, it proved to the skeptics the stability of the footing and tower. Indeed, it was a tremorous wink from the gods, a small, wry gesture announcing that Man and Nature shared in the business of the Golden Gate, and that both were tough enough to endure.■

Left
The pier surface was planed smooth, and adjoining slabs of 5-inch-thick steel were set down as foundations for the tower legs. They were secured with steel dowels, 6½ inches in diameter, driven into holes predrilled in both concrete and steel.

Right
Each tower was built as two legs, or shafts, joined by X-bracing below the roadway and by portal struts above. Here, with the initial sections already in place, Bethlehem steelworkers are preparing the base plates of the south tower.

Left
Early in the construction of the north tower, many workers were stricken with a debilitating sickness. At first a mystery, its cause was traced to poisonous fumes emitted when smoking rivets touched the red-lead-treated steel. Thereafter, men working inside the cells were required to wear respirator masks.

Right
We had to report for work at the base of the tower. We got paid from the minute we started work, not from the minute we climbed the tower—that we did on our own time. Usually it took 30 minutes to climb the tower and get ready for work. Except on Mondays—then it took 40.
(Francis Baptiste)

For raising the tower steel, a "traveller" was fit between the shafts. Comprised of two large steel trusses, it was equipped with a stiffleg derrick and a 90-foot boom to service each shaft. The traveller was designed to climb the tower in jumps of 35 feet. It was hoisted by block-and-tackle rigs and held in place by support posts built into the tower legs.

At the pier, a stiffleg derrick with a 100-foot boom unloaded the derricks and stacked the steel. As soon as the tower had risen 200 feet, connections between the bottom cells and the anchoring steel were made. Only under that much stress did the angle irons reach the engineers' specification for tension—105,000 pounds per angle.

Left
On May 4, 1934, the Marin tower was topped off, and an American flag was raised in salute. The men attending this informal ceremony stood atop the tallest bridge tower ever constructed. At 690 feet above the pier, it was 125 feet higher than the towers of New York's George Washington span, and 232 feet higher than those of the San Francisco-Oakland Bay Bridge.

Right
As yet unstabilized by the cables and roadway, the narrow and flexible Marin tower could sway menacingly in the winds. "Sometimes that damn tower would be snapping," recalls one steelworker, "and we'd just lay on our bellies and watch the San Francisco skyline go bobbing up and down."

At first, the tower top surface was little more than open girders and exposed rivets—at best a precarious footing. To prepare the area for the bustling activities of cable-spinning, teams of welders, carpenters and riggers constructed a spacious and secure workstage.

Left
When we was raising the tower, the connectors up high would wave hand directions to the signalman down below, who'd send light signals to the crane operator. Say we wanted a 65-ton cell to come down easy —that might be ding, ding, ding on the red light, and maybe leave the third ding on. And when he gets one flash like that, the operator lets the whole thing go, and that's how we dropped those big babies into place. (Peanuts Coble)

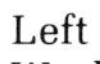

Left
Working inside the towers, you're all the time fighting the confinement, the darkness, the noise and the fumes. At lunchtime, if we was near the top, we'd grab our lunch-pails and go outside to eat. That'd give us at least an hour of fresh air. (George Albin)

Right
The tower shafts were set 90 feet apart. For the 200 feet from the pier to the roadway, they were joined by X-bracing. At the roadway level, sidewalk brackets were installed around the perimeters of each shaft. Men and equipment moved vertically between the shafts by way of a primitive open-air, block-and-tackle hoist called a "skip box."

Conventional bridge design called for X-bracing along the entire height of the towers. But on the Golden Gate, design engineer Clifford Paine and architect Irving Morrow made innovative use of portal struts above the roadway level. Paine found them structurally sound, and both men felt they were appealing to the eye.

With the Marin tower as a model, Bethlehem turned the Fort Point wharf into a welter of activity. *A lot of times we'd have to move up 20 feet from one connection to the next. Usually, they installed ladders as the tower went up. But who the hell had time to wait for the ladders? There we were with all our tools on—an 8-pound sledge, a sleever bar and a couple of spud wrenches—and we'd be climbing up the inside of those tower cells like monkeys up the side of a palm tree—stepping up the rows of rivet heads.* (Peanuts Coble)

Originally, the schedule called for Bethlehem to begin south tower construction immediately after it had completed the Marin tower, using the same crews and equipment for both jobs. But setbacks in the south pier project delayed the tower sequence nearly seven months. Not until January 1935 did the south tower begin to rise.

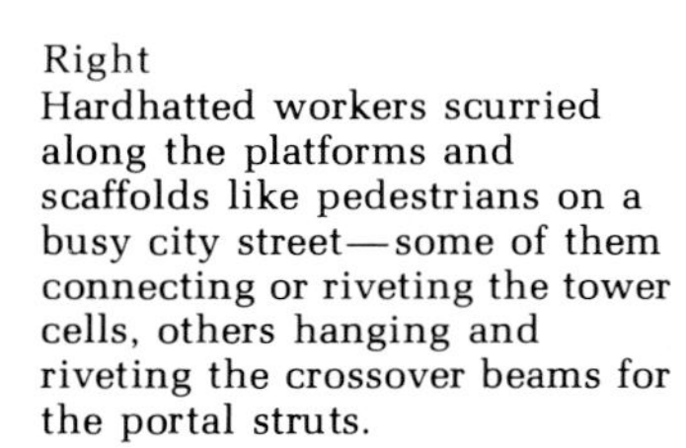

Right
Hardhatted workers scurried along the platforms and scaffolds like pedestrians on a busy city street—some of them connecting or riveting the tower cells, others hanging and riveting the crossover beams for the portal struts.

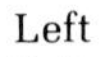

Left
Refined by the experience of the north tower, and responding to an urgent deadline, Bethlehem's crews raised the San Francisco side quickly. By mid-March of 1935, the tower was nearly 500 feet off the water.

Right
In the cramped enclosure of the tower cells, the riveters endured noise, discomfort and a pitiless production quota. "We riveted eight hours a day, 350 rivets a shift," recalls one Bethlehem worker. "If you couldn't make your 350, you didn't stay."

On top of the towers, Roebling's people installed the cable saddles, bringing them up in three sections weighing more than 500 tons each, then bolting them together to form the single saddle.

The saddles would nest the bridge cables as they were being spun across the gorge. The bottom of these weighty fixtures are grooved to accommodate the individual strands of cable.

The abrasive salt air of the channel could quickly corrode the tower steel, and from 1934 on, paint crews were sent out to the bridge. Their systems were primitive. They painted exclusively by brush, standing on plank scaffolds, or sometimes dangling precariously from bosun's chairs made from manila hemp. They looked like mountain climbers ascending a sheer cliff as they followed the weather, treating the scars on the steel.

A completed tower was a myriad of steel cells held together by 600,000 field driven rivets—honeycombed steel chambers built in so labyrinthian a pattern that men could get lost just climbing from top to bottom. *Finding your way out at quitting time was always a problem. One set of cells goes down for a while, but to continue you have to turn off and walk around to one of four openings —you never knew which was right—before you could find the extension of the ladder. Pretty soon you have no idea where you are. One day, a buddy and I made the trip from top to bottom —twice. It took us from 4 o'clock until 6 o'clock just to get out. And then it was only an accident that we ran across the right cell.* (Whitey Pennala)

Cables: The Gate Is Spanned

Many days saw 1,000 miles of cable placed in 8 hours of working time.
—Clifford Paine

Ropes Across the Water

The main cables of a suspension span give it visual identity. They drape the towers and link the shores, sketching the bridge in silhouette across the sky while holding the towers secure against the elements and stabilizing an entire modern highway. To these common duties the Golden Gate project added its own challenges. It demanded of contractor John A. Roebling & Sons main cables of unprecedented size and weight spun at record speeds.

According to Roebling's design, the cables would anchor at the shores, climb gradually the 1,125 feet to the near tower, span the 4,200 feet between the spires in a sweeping arc, then glide another 1,125 feet to the far anchorage. Each cable would be an amalgam of 27,000 rows of wire strung across the gorge by Roebling's unique on-site spinning method. Each would measure 36½ inches in diameter and 7,650 feet in length, and would weigh 7,125 tons. In all, they would require nearly 80,000 miles of wire, enough to girdle the earth at its equator three times over.

Roebling started the prodigious undertaking early, about the time the first of the tower steel was being erected. The company was headquartered in Trenton, New Jersey, but for a job of the magnitude of the Golden Gate Bridge it needed facilities much nearer the site. In the summer of 1933, Roebling purchased nine acres of bayfront land around a deep water cove north of Sausalito. There it built a subsidiary plant, staffed it with supervisors imported from the East, hired laborers and carpenters from among the locally unemployed, and went to work.

Beginning that fall, the bridge wire was shipped west from Trenton in 4,000-foot coils. It was unloaded at California City, steam cleaned of protective wax, rewound on larger reels, and warehoused. The cold-drawn, double-galvanized acid steel wire was remarkable material. It was thin—about the circumference of a number two pencil—and flexible enough to permit a 475-foot catenary sag from the tower tops to midspan. Yet it was so strong that even the toughest bridgemen could not bend an 18-inch length of it by hand.

In late spring of 1935 Roebling's activities began to merge with Bethlehem's. As ironworkers were riveting the final few sections of the south spire, Roebling's carpenters were building platform decks around the tops of the towers.

I was just another Depression statistic who found work at California City. I worked around logging camps before, rigging trees, but nothing I ever done prepared me for those bridge towers. One day, I was minding my business, hooking on lumber, when my boss told me to go out to the Marin tower and help the carpenters.

I went out there the next day, not really knowing what to expect. This was before they put any man hoists in, and with all our tools and ropes me and the carpenters climbed all the way up by ladder. When you first reach the top, you find that you're out on 750 feet of steel. You notice there's not much walking-around room. And then you notice the tower is swaying in the wind.

Well, we didn't quite make it that day, these carpenters and myself. Not one of us had the guts to stay up there and work. We were trusted guys—nobody was going to come out and check on us, and we knew it. So we just looked at each other, shrugged our shoulders, and climbed back down. We took the rest of the day off.

The next day we went back up with great determination and stayed. Eventually we even got used to it, although I don't know how. I understand from instrument readings that the towers moved regularly up to six feet. I don't know about that. But I do know it swayed noticeably while we worked, and that even experienced steel workers got seasick from time to time. (Bunk Mersereau)

In early suspension spans it was common practice to sway the towers back from the center before the cables were spun. Then, as the bridge slowly took form, the towers were drawn into a final, erect posture. The physics of the Golden Gate Bridge precluded such a plan, so Roebling engineers devised a roller system.

Once the platform decks were completed, the usual huge steel "saddles," weighing 160 tons apiece, were derricked up and put in position for nesting spun cable. But they were placed on a new kind of foundation plate. Each of these plates was a strong steel slab resting on thirty-four horizontal rods that worked like roller bearings. The plates permitted the saddles to move as much as 5.5 feet toward the channel on the San Francisco side and 3.7 feet on the Marin side, distances computed to offset any undue stress upon the spines of the Bridge.

With the bolting down of the saddles in late July, the Golden Gate was ready to be spanned. At 8 o'clock the morning of August 2, a dank, foggy day that saw tumultuous waves come crashing through the gorge, one end of an inch-and-a-quarter wire rope was tied to the Marin anchorage. Battling strong tides and howling high winds, a barge moved slowly toward Fort Point, unreeling 5,000 feet of rope in its wake, taking one full hour to cross the channel.

For the next two hours lines were sent down from the towers and hooked onto the rope. Then, while a barricade of Coast Guard vessels held back all ships, the derrick motors groaned, and the rope emerged glistening from the Bay, rising slowly to the tower tops.

This unprecedented shore-to-shore link was the first of twenty-five ropes, all hoisted by August 25, that formed the foundation for Roebling's catwalk work stagings. When finished, these footbridges would be 15 feet wide and suspended across the channel about 3 feet below the projected line of the main cables. They were built from redwood one-by-fours bolted crossways to light steel runners. The runners were 10 feet long and grooved to fit the ropes like train wheels on a track. One set of runners and planks formed a section; ten sections formed a 100-foot "train," the basic unit of footbridge installation.

For the backspans, the catwalk sections were joined into trains at ground level, then pulled up the footbridge ropes by derrick. Crews of carpenters literally rode them into position, clamped them to the foundation ropes with U-bolts, then returned to the ground and repeated the process. That was safe duty compared to the method being used on the mainspan.

There the sections were lifted one at a time to the lofty tower decks and coupled into trains. Then they were set on the catwalk ropes and nudged by derrick until their own momentum sent them sliding toward centerspan. Clamping crews followed in a construction car, a rectangular box that traveled on the ropes, its channeled bottom rattling along the lines like a gondola car scuddying sideways.

As on the backspan, clamping involved shoving U-bolts through the planks and runners and fastening them to the ropes. But here, at times, the men had to straddle one rope and balance them-

In the dour fog of a San Francisco summer, Roebling workers prepared for the cable operation, rigging the wire ropes that would span the Gate as foundations for the catwalk workstages.

On August 2, 1935, Coast Guard vessels stopped all ship traffic while a McClintic-Marshall barge brought a 5,000 foot length of wire rope from the Marin anchorage to the San Francisco anchorage. It would be derricked up across the tower tops and serve as the first of 25 foundation ropes raised in the ensuing two weeks.

selves by holding a second rope with one hand. With their free hands they would attach U-bolts onto a section of runners and planks.

In the course of constructing those footbridges we had some hair-raising moments. You know, the wind could get the ropes all fouled up with the sections of planks. One time it made such a mess we couldn't get to it with a construction car. Then somebody got an idea, and a couple of men volunteered for duty. They laid a plank across that set of footbridge ropes and U-bolted it down so it wouldn't slip. Then they laid themselves across that plank like it was a life belt, and they pulled themselves hand over hand down the ropes. They actually rested their bellies on that one-by-four plank and swam down the ropes, clearing them by hand as they went along.

It was as foggy as the devil that day. Those men disappeared from sight almost immediately. They were gone the better part of a day, and all we could do was wait and hope they made it back all right. (Walter Weber)

As the footbridges took form, cross spans connecting oceanside and bayside catwalks were erected, one at midspan and one each at the quarterspans. They were the first stabilizers for the stagings, which until then would yield without resistance to the moods of the wind. Later handrails were strung to make the walks even more secure. And for protection from the gusts eddying underneath, storm cables were attached, stretching in triangular webs from the tower bases to various points along the catwalks. In deference to shipping, the storm cables were equipped with warning lights. The effect at night was spectacular—Christmas colors seemingly suspended in the air, crossing the channel in intermittent winks of red and green.

By the end of September the hand rails were in, and the last few planks were being clamped in place.

This was about the time Harold Ickes, Secretary of the Interior under Franklin D. Roosevelt, was beginning to take an interest in the Bridge.

I was handling publicity then, getting pictures and stories of the Bridge to national magazines and the like. I knew Ickes was travelling down near San Diego with Roosevelt. For publicity reasons, I asked him to come up and inspect the Bridge. He agreed, but at the last minute he sent a message begging off, asking that his secretary—a man whose name I can't remember—be allowed to represent him.

Well, that was fine. The secretary arrived on a foggy morning that just happened to be the day Roebling people were completing the catwalk. We went out to the site, and the workmen told me we were just in time to get some pictures of the first actual spanning of the Gate. So we marched up the catwalk from the south end. The last piece had just been laid between the center of the Bridge and the north tower.

When we got to the spot it was blanketed completely by fog. But we could see where they had put down the final plank. I shoved the secretary across it ahead of me and told him he was the first man ever to walk across the Golden Gate. He became famous for that. (Ted Huggins)

By 1935 on-site cable spinning had been practiced in the United States for nearly three-quarters of a century. From the beginning it was a Roebling enterprise. The company was founded in 1848 by Prussian immigrant John Augustus Roebling, probably the foremost figure in the history of domestic bridge building.

Roebling's understanding of the suspension principle was decades ahead of its time. While other engineers were skeptical or even ignorant of the form, Roebling was stringing suspension bridges successfully throughout the eastern United States. In 1844 he designed and supervised the building of an aqueduct across the Allegheny River. In 1848 he constructed a general traffic span over the Monongahela. In 1855 he engineered the world's first suspension bridge for rail traffic, across the Niagara River gorge. He erected a succession of "world's longest bridges," the Allegheny River span in 1860, the Cincinnati-Covington in 1866, and his masterpiece—finished in 1883, fourteen years after his death—the Brooklyn Bridge, 1,600 feet between the towers, the prototype for all long-distance suspension bridges built thereafter in America.

Roebling was the first to master the proper ratio of strength to rigidity, the first to sink cable-securing eyebars into masonry anchorages, and, most significant, the first to adopt a method of on-site cable spinning originated in 1829 by a Frenchman named Vicat. Building a small bridge across the Rhone River, Vicat constructed cables in place instead of following the usual practice of spinning them on land then hoisting them into position across the towers.

The weakness of the system he replaced was obvious: it was impossible to fabricate great lengths of cable then raise them by machine. With the new system, which Roebling would call "parallel wire construction," it would theoretically be possible to form cables of infinite length and diameter using a coil of reed-thin wire and a rope. In applying this principle successfully, J. A. Roebling revolutionized the science of suspension bridge building.

Though difficult for a layman to visualize, parallel wire construction follows some simple concepts. It involves laying reeds of wire in rows, one on top of the other, until a proper girth has been reached. The process resembles in reverse the winding of a skein of yarn into a ball: cable-spinning takes the ball and unwinds it into a skein. A reel of wire sits just behind the anchorage on a bobbin in an unwinding machine. The loose end of the wire is drawn out a short distance, then looped back and tied to a horseshoe-shaped fixture called a "strand shoe" which is imbedded in the anchorage. The loop is hooked around a pulley wheel (or spinning "sheave") that hangs from a carriage mechanism on an overhead tram. The machine-driven tramway line pulls the spinning sheave from the anchorage, over the towers, to the far side of the bridge. There the loop is taken from the wheel and wrapped around the corresponding strand shoe on that side of the span. Then a new loop is formed and hooked around the pulley wheel, and the carriage is cranked up for a return trip. In this fashion, two wheels (one for each main cable) run back and forth laying wire.

A set number of wires form a "strand" (on the Golden Gate it varied from 256 to 462). When that number has been reached, the strand is tied off, banded, and jacked into position across the towers to maintain the proper sag. The operation then moves to another set of strand shoes and begins anew. A predetermined number of strands compressed together form a bridge cable (it was sixty-one for each cable on the Golden Gate). When all the strands are spun, they are compacted into a perfect circle under severe hydraulic pressure. Then they are wrapped with a layer of fine wire. The massive cylindrical cable now runs between the anchorages from splay point to splay point—

those points where the wire strands fan out individually to meet the strand shoes.

This remarkable procedure put the Roebling name on the masthead plaques of America's most significant bridges: the Williamsburg in 1903 and the Manhattan in 1908, both crossing New York's East River; the Bear Mountain near West Point, New York in 1924; the gargantuan George Washington Bridge crossing the Hudson River in 1931. It was the procedure that Roebling brought to the Golden Gate Bridge in the fall of 1935.

Cable spinning is a finely calibrated series of delicate steps—reeling, splicing, measuring, surveying, adjusting, clamping, transferring, repairing, monitoring—precision work paced by the relentless speed of the hauling wheels, which on the Golden Gate could roll 650 feet per minute all day long.

In parallel wire construction, the bottom or "dead" wire of the spinning loop, the one attached to the strand shoe, is subject to constant adjustment. The other, the top or "live" wire, runs non-stop from the anchorage up over the towers until the return trip, when it becomes the dead wire for the next pass.

As the dead wire reaches a predesignated adjustment point, it is tied to an electrically powered "comealong" rope called a Selysun unit. This unit is replete with dials and counterweights, and it is operated remotely by an adjuster at midspan. Manipulating controls, he can pull up the wire or slack it off, to bring it to an exact tension. When it is dead-center perfect, he sends a light signal to a clamp-off man who ties down the wire. All this occurs while the live wire is still running toward its anchorage strand shoe, allowing the cable to be tuned precisely while it is being spun.

Within days after the footbridges were finished, Roebling was educating its new men on these intricacies of on-site cable spinning. It wanted only trained mechanics who knew exactly what to expect.

Roebling had quite a crew of supervisors they brought out from the East—men who really understood the work. And for the rest of us, well, they hired and fired and hired and fired until they shook down to just the kind of crews they wanted.

Before they let any of us get near that spinning operation, they sent us down to Fort Point for a week-long training session. They instructed us on the individual duties we were expected to perform. They only taught us our one job; they wanted us to do one thing to perfection, not a lot of things poorly.

At the end of the week, Mortenson, the foreman, said, "Men, you've been hired for the duration of the job. We know you can put out, or we wouldn't have kept you. If there's any trouble or tie-up while you're working, we have certain crews who know how to fix it. The rest of you don't worry; you don't have to run around trying to find something to do. If you want to sleep, sleep. We know what we want you to do; we know when we want you to do it; we know you can do it. So when there is work to do, we expect it done. When there isn't work, you are on your own."

I had never worked for anybody who had that attitude. Most companies would worry about losing money if everybody wasn't working all day every day. But Roebling had class. They felt their men were individuals, not cogs in a wheel. And they knew how to do a job. (Gerry Conser)

At the close of instruction, the men positioned themselves along the catwalks to prepare for spinning. Reeling machines were installed at the anchorages. A-frames made from pipes and ropes were constructed as support posts for the hauling trams. Four-walled booths were built as signal huts. Transfer points were equipped for the smooth exchange of loops. Warm-up shacks were erected at intervals for escape from the numbing wind.

Meanwhile, company engineers, the specialists from the East Coast, were busy surveying the projected sag of the cables.

We couldn't very well start spinning until we knew exactly where we wanted the wire to hang. And for that we had to put up guide lines. My job—I was a contractor's man—was to calculate precisely at what elevation we had to erect the cables so they would end up matching the engineer's dimensions. So before the spinning began we strung guide wires across the span and surveyed them into place. We did most of the work at night, so there would be no thermal expansion to distort our calculations. And from the time those wires went into place they were law—they acted as a measuring tape for all 54,000 wires that went into making up those main cables. (Walter Weber)

Roebling demanded fail-safe coordination between the stations along the catwalk. A wire jumping its trolley at midspan would affect all spinning in front and behind. An adjustment made at the tower had to be registered at quarterspan; crew members had to be reliably informed at all times. For that Roebling built in impressive communications back-up systems.

Every signal had a standby alternative. A man at an adjustment station unable to hear a ringing phone could be alerted by flashing lights. A man who could not make out lights through a thick fog could be reached by bells and horns. In all, Roebling electricians used more than 200 miles of copper conductor activating this dense network of devices.

Yeah, that was really a system, flashing lights, telephones and bells. But we needed all those things, we had that many signals to give. We had wires to watch and adjust and we had to let people on down the line know what was going on. The system was so good that if you made an error you couldn't lie your way out of it. When we first started spinning cable, and none of us knew too much, we'd naturally try to alibi any mistakes. But they had a dispatcher at midspan. Every time the spinning wheel stopped—I think it was costing Roebling 1,200 dollars a minute to run those wheels—it was their job to report the reason. And if it was a mistake that happened to be your fault, they could trace it right down to you. (Earl Maillioux)

While the men were at work on the spinning operation, the company brass were fretting over a different set of problems. On the George Washington Bridge Roebling had strung wire across the Hudson at record velocities and volumes, running the wheels up to 750 feet per minute, spinning up to 61 tons of wire a day. Those were impressive figures in 1931, but four years later they were inadequate.

If Roebling just equalled that output on the Golden Gate it would finish the project devastatingly in the red. Business competition was brutal in the Depression, and to win the Gate contract, Roebling had trimmed all the fat from its bid. The result was a 5.8 million dollar job that threatened to lose money for the company unless it could be executed in fourteen months.

At first reckoning, that seemed an impossible deadline. Deducting a reasonable number of days for constructing the footbridges, discounting time spent in breakdowns and weather delays, the time for actual spinning would be nine months. Given the dimension of the task, and the more than 1,400 tons of wire needed to form the main cables, the company would have to spin 265 tons a day, 204 more than existing methods would allow.

It was an awesome peak to scale, but no company was better equipped for the ascent. In fact, Roebling had experimented on the final strand of the Hudson River span. Up to that time, each tramway carriage pulled only one spinning sheave and one loop of wire. Roebling engineers reasoned that if a carriage could manage one wheel, it could also manage two. The idea worked well enough in limited practice. Roebling was convinced the Golden Gate trams could carry two wheels each, doubling the daily tonnage to 122.

That was just the beginning. In previous bridge erection the tramway carried the spinning sheaves from anchorage to anchorage and back again. For the Golden Gate, Roebling designed a "split tram." In this new setup, there would be two hauling carriages for each cable, not one. They would originate at opposite ends of the bridge and travel toward each other on the same track seemingly on a collision course.

At midspan, the trams would be halted, and the loops transferred; two wheels travelling halfway then returning could produce the output of four single wheels. That pushed the daily output in theory up to 244 tons.

These innovations would work in normal weather, but the spinning was to begin in November, the month that winter launches its annual five-month assault. Roebling's people learned quickly the effects of a Pacific north coast stormy season. Conditions on this bridge, many claimed, were the worst the company had ever encountered. The first few weeks were expected to be sluggish; it would take that long for the crews to grow confident and sure. But on top of that the blustery wind and rain delayed work for days at a time, spreading doubt over the gains won so optimistically on paper.

There's two kinds of freezing in bridge work. One is where you're so scared you can't move a muscle. That happened to me once, when I went across the walkway and hit a wire and turned my foot. It literally froze me:

I couldn't move a muscle. A steelworker was up above, and he saw me. He started calling me names. That's an old trick—you get your mind off being scared by getting mad at him instead. But the real cold was a bigger problem. We wore three sweaters up on those catwalks and still couldn't keep warm. We got to wearing our raincoats just to ward off the wind, that was on the days we could work at all. (Francis Baptiste)

There were schedule setbacks, and Roebling engineers were stumped initially for a solution. Finally, in January, 1936, with a third of the 122 strands already spun, Charles Sutherland of the San Francisco field office thought of an answer: if the hauling trams could accommodate two wheels each, why not three? In theory there was nothing to prevent three wheels on each of four trams from running twenty-four wires simultaneously.

Sutherland received approval from Roebling supervisor Bob Cole, and together they brought the idea to the field. There, workers already under extreme pressure spent weekends installing a third wheel and reeling machine for each carriage. The three wheels hung from the hauling trams like the interlocking hoops of an Olympic logo. They looked odd, and at first they confused the workers. Men used to monitoring first two wires, then four, now had to keep track of six.

Roebling helped solve the problem by color coding the wires as they came off the reels, and the men responded fast. They needed a week or so to become adept at handling more wire, but once that passed, work along the footbridges accelerated. The times recorded were impressive. At peak output, the wheels would travel from anchorage to midspan in just 6½ minutes. In one minute more the loops were exchanged, and the wheels were rolling again. Output capacity was pushed to an unheard-of 271 tons of wire a day.

The duties along the catwalk ranged widely: an electrician might be checking transformers one day and replacing tower-top light bulbs the next. A laborer might be stationed in a signal hut, or he might be assigned a spot just to watch for snags.

. . . ever see a wire get a kink in it? Out on the bridge we called that an asshole. Every once in a while the wire would develop a asshole, and the guy that was watching for them would stick a piece of wood into the loop and flip it back around. (Pete Williamson)

But mostly the men were situated to maintain the progress of the wheels. They would watch the unreeling machines and splice the end of one spool with the beginning of the next. This they would do with small steel turnbuckles that were double-threaded, left on one end, right on the other. The men would thread the wire ends and ratchet the turnbuckle tightly into place, the same procedure they would follow should a wire break out on the span. There were also men to work the Selysun units, men to change loops at the transfer stations, and men to circuit-ride the catwalks, lubricating pulleys and staying alert for trouble.

The wheels hummed through the first months of 1936. In mid-January, twenty-two strands were in place around the anchorage shoes, thirteen on the Bay side, and nine on the ocean side. By late February, twenty-six more were completed. There was no reason now to doubt the amended spinning system. The only question was by how much the cables would beat the schedule.

As finished strands were jacked into position across the towers that spring, bets were placed on the completion date for the 122nd strand. The pace was picking up. In the second week of April alone, eight strands were spun and positioned. That left only twenty-four, twelve over each catwalk, and the betting pots grew larger. A timekeeper named Paul Zeh guessed 2:15 PM on the 20th of May and put his money down. He missed by two minutes. At 2:13 on May 20, the final strand was set into place across the ocean-side tower saddle.

The Roebling crews had taken only six months and nine days to complete the mammoth job. They chopped months off an already difficult schedule by rolling wire sometimes 444 per cent faster than had been done on the George Washington Bridge. To celebrate, the company threw a prolonged party at Paradise Park near California City.

For three days raucous sounds rang through the Tiburon Peninsula, and the reasons for them were clear. These were hard times, but both management and labor had earned big money. And by their historic efforts the bridge project was back on schedule. There was still work left on the Roebling contract—cables to compact and wrap, bands to install, suspender ropes to hang—but the big job was finished. A bridge was in silhouette across the channel, and the Golden Gate was just a floor system away from being a completed span. ■

Up on the tower saddles, Roebling ironworkers helped arrange and secure the catwalk ropes as they were being derricked up the 746 feet from the channel waters.

Where they rose to cross the tower tops, the footbridge ropes passed through turnbuckles which were used to adjust the ropes for slack.

Left
Catwalk construction often involved hazardous duty. Many times Roebling ironworkers performed their jobs several hundred feet above the water, supported only by the thin circumference of the foot-bridge ropes.

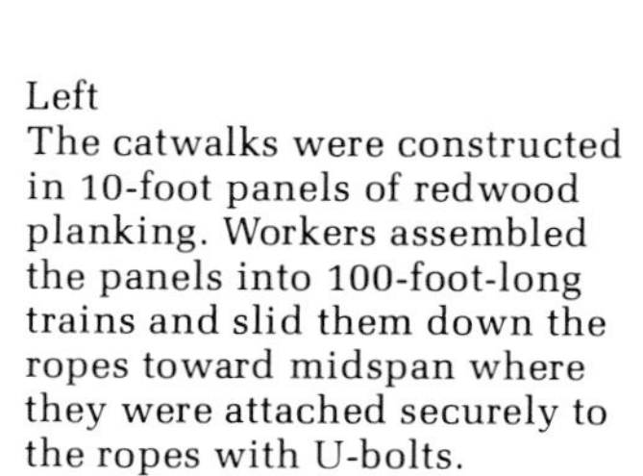

Left
The catwalks were constructed in 10-foot panels of redwood planking. Workers assembled the panels into 100-foot-long trains and slid them down the ropes toward midspan where they were attached securely to the ropes with U-bolts.

Right
The footbridge ropes swooped across the Gate in a catenary curve approximately three feet below the projected line of the main cables. When all ropes were in place, workers stablized them by installing a crosswalk platform at midspan.

By late September of 1935, the catwalks were intact. Crosswalks had been installed at quarterspans and midspan; wire rope handrails were strung the entire distance—a person could now walk from San Francisco to Marin along this redwood plank footbridge.

To escalate the workers along the steep incline to the tower tops, a machine-driven hand trolley was installed on the oceanside catwalk. And to stabilize the walks themselves, a set of storm cables was strung in triangular webs from the bases of the towers to the bottoms of the walks.

By October of 1935, the catwalks were rigged for cable spinning. Overhead tramway lines were hanging from anchorage to anchorage, and along the tram, generated by machine, rode the spinning wheels. Also called "sheaves," the wheels would pick up loops of wire at one anchorage and carry them to center span for transfer and continuation to the opposite end. Nearly 80,000 miles of wire would span the Gate in this manner; for each cable that meant some 27,000 individual lengths of wire piled into 61 bundles, called "strands." Spinning that much wire across a wide and wind-blown channel was imposing enough, but the specter of profit was also hovering about the job. Roebling had to execute the contract by July of 1936 just to finish in the black.

Left
At the anchorages, each strand of bridge wire looped around a specially wrought, horseshoe-shaped fixture connected to an eyebar called a "strand shoe."

Right
A closeup of the Marin anchorage shows a strand shoe coupled to an eyebar, with a completed strand of wire looped around the pulley groove of the shoe.

Left
The spinning wheels measured approximately four feet in diameter, and it required a crew of men to fit one of them with the "bights" (loops) of wire.

Right
At first, the spinning was performed with just one wheel per catwalk, each wheel carrying a single loop of wire—the standard arrangement in suspension bridge construction. But that system proved unprofitably slow, so Roebling added a wheel to each tramway, thus doubling its output.

Left
After the strands had been spun across the tower saddles, the men adjusted them for tension and sag with huge hydraulic jacks. Here, the workers are installing the lifting straps and seizing bands that would serve as handles for the jacks.

Right
We had warmup shacks out on that catwalk—God we needed them. The wind blowing in from the ocean forced you to stand at 20 or 30 degree angles all the time, and the dampness and the cold cut right to the bone. Those catwalks were the coldest goddamned place I ever worked, but Roebling saw to it that we were all relieved regularly so we could go inside for a few minutes. (Pete Williamson)

At a midspan spinning station, workers alertly watched the progress of the wire. When a strand was formed—the number of wires in a strand varied, but the average was 452—they banded it tightly by wrapping it at five-foot intervals with several turns of heavy adhesive tape.

In January of 1936, with 42 of the 122 strands already spun, Roebling increased the output by adding a third wheel to each of the hauling trams. The triangular rig on the catwalk was a "separator." It kept the individual strands segregated by the width of the steel rods that hung like harp strings from the wooden frame.

50 YEARS:
GOLDEN ANNIVERSARY PORTFOLIO

This portfolio is published in celebration of the 50th Anniversary of the opening of the Golden Gate Bridge. These historic photographs were taken in 1936 and 1937, during the final year of construction, and were originally printed in black and white. They have been hand-tinted for this special occasion and were chosen to recall the exciting, triumphant completion of what would become the most famous, most admired and certainly best-loved bridge in the world.

A view from the Bridge.

The roadway is literally suspended from the main cables.

Roadway paving will soon hide the
steel grid work.

Looking north across Fort Point.

Compressing the strands of the main cable.

"Will it meet in the middle?" was a popular question.

The unique safety net is credited
with saving 19 lives.

International Orange, the final coat.

The Golden Gate is visibly spanned.

Opening Day. May 27, 1937.

Left
For sculpting the 61 cable strands into final form, Roebling employed hydraulic compacting machines that were equipped with a circumference of radial jacks. Capable of applying 4,500 pounds of pressure per square inch, they left in their wakes two perfectly round bridge cables which measured 36 1/16 inches in diameter.

Right
With the cables compacted, Roebling's steelworkers began bolting up the cable bands, the huge cast steel clamps they situated every 50 feet for hanging the suspender ropes.

Left
The completed strands sat in a natural hexagon, and to form them into a circle, the radial jacks of the compacting machinės clamped like teeth around their full girth. The tiny protrusion in front of the jacks is a "ferrule," used to splice wire ends from one reel to the next.

Right
The suspender ropes were installed by agile steelworkers who often nudged them into place while perched precariously on top of the main cables. The ropes were lowered to the workers from the hauling trams, then fitted into position and allowed to dangle like kite tails for weeks until the roadway steel was raised.

Left
At the tiedown points near the pylons, multiple cable bands holding some of the span's shorter ropes sent 8 loops, or 16 lines, dangling toward the roadway.

Right
A standard cable band served as a fixture for two loops of rope, which sent four lines, hanging in a square, down to the roadway. The length of the ropes varied with the catenary curve of the main cables; those nearest midspan and the endspans were only a few feet long. Those nearest the towers were 980 feet in length.

Left
After the cable bands were installed, Roebling applied the finish coat. Six wrapping machines rode the cables, winding a layer of fine wire around the circumference of each cable. With the wrapping, the cables appear to be covered by thin sheet steel.

Right
Roebling had a fire pressure system on top of the towers, with one-inch pipes fitted with outlets running the length of the catwalks. One night, it got so cold out on the Gate, the pipes froze up and broke, and the leaking water turned into icicles. Until it thawed, we couldn't work—everything was just a frozen mass, with icicles spreading out in every direction the water had leaked.
(Gerry Conser)

By the time San Francisco's summer fog rolled in for the year 1936, the main cables of the Bridge were completely intact. In record time, J. A. Roebling & Sons had strung the world's longest and thickest bridge cables across one of Nature's deepest and most belligerent channels of water.

Just below the pylons were the "splay points," where the individual strands fanned out from the main cable to meet the strand shoes at the anchorages. The splay points were exposed only for a short time—Barrett & Hilp returned to construct walls and ceilings around the entire area, in effect building an "anchorage" house.

By mid-summer of 1936, the Golden Gate channel was visibly spanned. Lofty towers were anchored to immovable piers. Thick and pliant cables were hanging in graceful arcs from shore to shore. The once-resistant gorge might still snort and fume, but it had been tamed. All that remained were the final large steps: the raising of the floor system steel, and the pouring of the highway. In less than one year the Bridge would be open.

Roadway: The Gap Is Closed

Every span is something that "can't be done" until the men in steel helmets have driven in their last rivet.
—Joseph Strauss

Walking the High Steel

Architect Irving Morrow had been part of Strauss and Paine's staff since 1930. His selection had surprised the more staid members of the architectural community. Morrow had no particular standing in the field. His only credential was a large modern house he designed in the Forest Hills section of San Francisco, and he had earned the enmity of traditionalists with open criticism of San Francisco's inclination toward Gothic and Renaissance design.

But, like Strauss and Paine, Morrow was a modernist before the term was widely accepted. He believed that architecture should speak only for the age in which it lives. What's more, he agreed with Paine that the Golden Gate Bridge be functionally designed.

In most ways, Morrow's job was dictated by the dimensions of the span—the height of the towers, the width of the roadway, the sweeping arc of the cables were all matters of applied physics. Nonetheless he exerted an important influence upon the look of the Bridge.

He convinced Paine that the portal bracings above the roadway deck should diminish in size as they climbed. He introduced fluted cover plates and corner brackets for the tower portals and fluted edges for the concrete piers and pylons. For benefit of sightseeing motorists, he avoided the traditional view-blocking sidewalk fence by designing instead a railing of open balusters.

Morrow's most striking contributions were his decisions on color and lighting. He imagined the towers illuminated as one sweep from water to sky, and the roadway as a mellow glow that would lead the eye directly but softly across the channel. So he designed lighting that would not call attention to itself, occasional globes suggesting only the lines of the span. And to assure that the roadway would not be obscured by frequent fogs, he insisted on expensive sodium vapor lamps that had been just recently invented.

He proceeded just as thoughtfully on the matter of color. In the spring of 1935, he authorized several paint companies to recommend formulas and test them in the field. This they did on small panels which were exposed to the weather on the roof of Fort Winfield Scott. By April of 1936, the choices had narrowed to carbon black, steel gray, and orange-vermillion, a tone similar to the red lead that was already protecting the steel.

The strongest advocate of either black or gray was O. A. Ammann, who thought the gray on his George Washington Bridge was a most serviceable color. Morrow agreed, but believed it too dull for the Golden Gate. He wanted a warm color to contrast with the cool tones of Nature, the grays, blues, and greens of the water, sky, and clouds; and he wanted a luminous color to enhance the scale of the span. Thus the decision for orange-vermillion—officially called "International Orange"—a choice applauded by San Francisco artist Beniamino Bufano.

"I have been watching closely the progress of the towers of the Golden Gate Bridge," said Bufano in a letter to the Building Committee; ". . . let me hope that the color will remain the red terra cotta because it adds to the structural grace and because it adds to the great beauty and color symphony of the hills."

Once International Orange was chosen, the Bridge painters began using it to replace the red lead undercoating they had been applying all along. A ten-man crew had been continuously painting and repainting the newly erected steel since the towers started to rise in 1934. They blended with each phase of construction, working from bosun's chairs that dangled from portable scaffolds. They moved about the span as the weather allowed, painting first red lead, then International Orange, in a never-ending and unpredictable sequence.

We never got caught up on that damn Bridge. Some days we couldn't paint outside because the fog made the steel so wet. Then we'd work inside the towers. When we'd come out again, we'd always run across work we'd done recently that was so weather-beaten it had to be painted again right then. We criss-crossed that job so many times before and after the Bridge was open that I couldn't tell you how many times it was painted, and I worked on it continuously from 1934 to the day I retired in 1963. (Jimmy Biondi)

Only with the laying of the roadway does a suspension bridge come fully alive. Up to then the towers are idle self-sustaining units holding up the main cables like wooden poles supporting a clothesline. But when the roadway steel is raised, the towers and cables find vitality and purpose.

The process begins with cable bands, huge cast steel clamps forged in diametrical halves. They are hoisted to the tower tops by crane, then lowered into position by carriages which ride the tramway support ropes. Designed to fit flush around the main cables every 50 feet, these bands are bolted into place by a machine-driven wheel wrench, 5 feet in diameter, applying 120,000 pounds of pressure.

Steelworkers drape two loops of 2 11/16 wire rope around each band so that four lines dangle from the cables to the floor system steel. These are the suspender ropes which will hold up the roadway. They fit into sockets bored through the outer flanges of the main longitudinal girders, and are ratcheted into the floor steel, giving the bridge its final, interdependent shape: foundations supporting towers supporting cables supporting a roadway.

In the summer of 1936, in an atmosphere of synchronized frenzy, the Golden Gate's floor system was ready to rise. Roebling was compacting the main cables to receive the bands. Bethlehem was rigging the towers for assembly of the erecting cranes. J. H. Pomeroy was preparing for construction of the steel arch over Fort Scott. All approach routes were underway. Bids were being taken for the building of the toll plaza.

The floor system was the last big step and the most perilous. Towers are built from inside cells, cables are spun along storm-secured catwalks, but roadway work exposes the bridgeman totally to the elements. Floor steel goes in as a latticework of connecting members which extend out from the towers in both directions. Beams and girders run horizontally, vertically, laterally, and diagonally, with little between them but air. To build a floor, bridgemen must connect and rivet while balancing themselves on 6½ to 8½-inch girder flanges. There are no handrails on the floor steel, no walkways, nothing but full and open view of the blue-green water foaming some 220 feet below.

When you're good—only then can you work up in the air like that. They call you aces then, and you're like fighters with fast footwork, quick hands and plenty of guts. You're right up there with the angels, and you're moving all the time. You work with your hands, but you're keeping balanced with your feet. You have to carry equipment too—wrenches, bolts, sledges—sometimes 30 to 40 pounds of tools. You had to have guts and skill to work high like we did on the Gate job. But I loved every minute of it. It made me feel good to know I was doing something that most guys couldn't. (Albert Zampa)

On the Golden Gate Bridge the risk of raising the roadway was heightened by the treachery of the channel itself. Malicious winds were apt to gust and swirl without warning. Sodden fogs could glaze the surface steel to an ice-slick footing. Mindful of these conditions, Strauss and Paine felt compelled anew to consider the issue of safety.

Already they had defied the odds. The neighboring San Francisco-Oakland Bay Bridge, six months from being open, had so far killed twenty-two men, a number considered acceptable by Timothy Reardon, head of the state's Industrial Accident Commission. By contrast, not a single life had been lost on the Golden Gate, owing in part to good luck, but mostly to the foresight of the engineers and their extraordinary precautions.

Since 1933 doctors and nurses were on call at a field hospital near Fort Point. Special diets were prescribed for high steel workers to counteract dizziness. Men with hangovers were required to drink down a cure of sauerkraut juice. To ward off "snowblindness" from the sun reflecting brightly off the Bay, tinted goggles were issued. The Golden Gate was not the first big job to feature hard hats and safety lines as some have claimed. But it was the first to enforce their use with the threat of dismissal. The twenty-second victim of the Bay Bridge neglected to tie himself in, slipped and fell. The Golden Gate allowed no such independence.

Made from 6-inch squares of ⅜-inch manila hemp, the safety net stretched ten feet wider than the roadway on either side and bellied down 60 feet. To secure it, the steelworkers tied its edges to wire rope, which they then threaded through specially built steel outriggers that were spaced 50 feet apart.

Yet as the Bridge was readied to receive its floor there remained a nagging fear that "zero deaths" was too much to ask of fate. Engineers had beaten the one-life-per-million-dollars figure before, but nobody had ever built a major bridge without some fatality. It seemed in the summer of 1936 that the Golden Gate was taunting the law of averages.

In June, Strauss and Paine unveiled probably the most dramatic safety measure in the history of bridge building. They ordered the manufacture of a huge trapeze net to be strung underneath the floor girders. Tied to outriggers extending from the sides of the steel, the 6-inch-square mesh would belly down about 60 feet below where the men were working. The purpose of the net was twofold. Strauss and Paine believed quite obviously that it would save lives. But they also felt that men performing without fear would work faster and more surely, thereby trimming costly days off the length of the job.

On both counts they were correct. The net cost only $130,000, including the price of the erecting frames. That figure was more than made up by the consequent speed of the steelworkers. They raised the floor in a little less than five months, putting up as much as 640 tons a day. And in the course of construction at least nineteen men fell into the net who would have otherwise dropped helpless and screaming toward that hideous death the bridgemen call "falling into the hole."

The net was built onto a large steel frame that was wider than the bridge. It moved out from the towers along with the traveling derricks, hanging by roller clamps from the lower flanges of the main girders. In Bethlehem's system the floor was erected in "panels," or sections, between the floor beams, the main lateral members of the roadway steel. As each panel of steel was completed, the net was sent forward and a new section was laced in from be-

Ordering a trapeze-type net placed under the roadway steel was probably the most dramatic safety measure in the history of civil engineering. Prior to this Strauss & Paine innovation, bridgemen worked under extreme pressure on narrow beams and struts, with nothing underneath but water.

hind. In this way it progressed just ahead of the raising gangs, eventually stretching all along the underbelly of the floor.

Floor work was scheduled to begin the third week in July, 1936 but an engineers' dispute with Roebling resulted in a delay. The contract had commenced in mid-June, and by July 20 the towers were rigged, the traveling derricks assembled, and the beginning few panels cantilevered in place. Then, when it was time for Roebling to raise the cable bands, the job stopped altogether.

Several bands from Roebling's first shipment had been tested at the University of California and found deficient. Clifford Paine conferred with his eastern inspector, a man named Baker, who claimed that Roebling's forging was inadequate. Paine, accepting his word, refused the entire shipment and demanded a new batch of bands.

Roebling complained vigorously, maintaining the faulty bands were exceptions to an otherwise satisfactory set. But Paine held fast, and Roebling had no choice but to relent. For the next six weeks Roebling manufactured the new parts and shipped them west. Not until August 31 were the steelworkers back in business installing the bands and hanging the suspender ropes. Shortly thereafter, on September 11, Bethlehem began its contract in earnest.

The job was basically a repeat of the tower construction. Steel pieces were barged from Alameda to storage platforms on the ocean side of the tower bases. From there they were hoisted to the deck by a stationary crane called a "Chicago boom." And, as with the tower sequence, the weather made no concession to man's convenience.

You talk about swells coming in. One day our stationary barge was bobbing up and down, I'd say about 15 or 20 feet. We tied ourselves in right with 2½ inch hawsers and some 1⅛ inch cables. But those swells kept rolling in so high that finally the barge came down hard enough to snap the lines, and here we go. The tide was running out the Gate at the time—it must have been about 4:30 in the afternoon—and it was like we were marooned on that barge. We had no motor, no way to steer it, and nothing to do but go with the tide. We were almost out the Gate, running toward the Potato Patch, before a tugboat caught up with us and threw us a line. I don't know how they even saw us. It was dark by that time, and we didn't have a light. (Joe Walton)

At the roadway level, the pieces of steel were carried to the raising gangs on flatcars that rode along the longitudinal floor girders, the stringers. Where the steel ended, the pieces were picked up by a traveling crane called a "stiffleg derrick" and hoisted into position for the connectors.

Actually four floors were being built at once. Each tower was outfitted with two traveling derricks, one for raising the steel between the tower and midspan, and the other for the backspan between the tower and the shore. Each derrick was fully manned with raising gangs, bolter-ups, and riveters, allowing construction to proceed simultaneously in four directions.

The first pieces to be connected were the top chords, huge longitudinal beams at either side of the roadway level. These hollow box girders measured 2½ feet around and were the main components of the floor system steel. Hanging from them vertically and diagonally were support pieces, to which were connected bottom chords, parallel and identical to the top chords.

These first connections formed the bayside and oceanside faces of the roadway steel, called "stiffening trusses." With their construction the floor system finally met the cables. The suspender ropes dangling from the cable bands were pulled into position and bolted securely into channeled pre-cast flanges attached to the top and bottom chords.

Then the lateral pieces went in. First the main members, 90-foot floor beams, were set inside and perpendicular to the chords every 25 feet like railroad ties. These beams were followed by stringers and sidewalk panels, which completed the unit of construction.

The bridgemen measured their work in panels, each panel being the set of connections between two floor beams, about 25 feet of steel construction. The raising gangs worked a panel at a time, gauging their progress by the number of panels they installed in an eight-hour shift. The riveters working behind the bolter-ups had to hustle to keep a specific number of panels between themselves and the raising gangs.

All that autumn the floor steel edged, panel by panel, from the towers toward midspan and the shores.

Work had to progress out from each tower so evenly that the cables were always holding up the same proportion of weight. Any riveting holes that didn't line up right couldn't be driven until one crew was aligned with all the others.

On the Sausalito end of the Bridge I was promoted eventually to rivet foreman. One of my jobs was to go out every evening to the points that wasn't yet riveted and check them with a pair of feelers. I had a given amount of space I could allow. We couldn't bring the riveting gang in to start driving until my feelers went in tight or wouldn't go at all.

One time they had a wreck out in central span, and it delayed erection for three days. All the raising gangs except for where I was had to quit work just to keep the whole construction in balance. There was a lot of men standing around. Bethlehem wanted them to keep busy, so they brought them over to our end. It was pure pandemonium—none of them wanted to work. They all wanted to stand around and talk. (Harold McClain)

The floor system was far easier to build than the towers. The net allayed even the strongest fears, and the men were able to work fast. The derricks moved horizontally along the girder rails rather than in leaps up the vertical inseams of the tower legs. Heaters threw rivets 30 or 40 feet to catchers they could see, rather than shooting them blindly 120 feet through pneumatic tubes. The air was fresh and the lighting was good.

All through the floor construction only a few jobs slowed the pace. Riveting in the chords, for one, was tedious and enclosed. These hollow girders were solid steel plate on three sides laced with diagonal strips on the fourth. To buck up a chord, a small man had to crawl inside and work in a 2½-foot space.

The chord sections were spliced together with great big plates, and that took a lot of rivets. To get at them, we'd just cut one of those lacings and hinge it backwards, out of the way, then one of us would crawl inside with a wildcat jam.

The only person who could buck up a chord was a little guy. So they went around and picked eight or ten of us small fellows and volunteered us for the job. I was in them chords for a long, long time that fall. And when the rivets would come in there through pneumatic tubes, the wind would gush in and blow rivet scales all over. That mixed up with the dust and dirt that swirled in all the time, and every shift my hair and skin would turn orange from the grit. (Earl Maillioux)

More than any other phase of the Bridge, the progress of the floor could be assessed by sight. Anybody within view of the span that fall of 1936 could see giant arms of steel growing daily in length. By the 21st of September, the floor extended 300 feet toward midspan from each tower, and 100 feet toward the shores. By October 18, it was 1,250 feet and 650 feet.

The last of the suspender ropes were hung on October 25, and by November 2, only 100 feet of open channel remained. Excitement ran high among the bridgemen. "We'd take the boat back to Fort Point at quitting time," remembers a steelworker. "The whole way we'd look back at the Bridge, pretty damn proud of the work we were doing. Then when we landed, we'd head for the pubs and start building it all over again."

On November 18 all of San Francisco's morning papers shouted the news from the Bridge. THE GAP ON THE GOLDEN GATE WILL BE CLOSED TODAY, read the *Chronicle*, the city's largest paper. The floor system was one chord from connecting the sea cliffs of San Francisco with the headlands of Marin; the joining piece was scheduled for afternoon installation. It had to be afternoon, because only then would the sun expand the steel sufficiently to allow a fit. Some 6,450 feet of interconnecting steel pieces formed the floor system, and they could meet in final junction only when the bridge was made malleable by the sun.

No elaborate ceremony ushered in that last piece of steel. How would celebrants have gained access to the site? Only a few civilians were on hand: George P. Anderson, president of the Redwood Empire Association; William Filmer, president of the District board of directors; E. B. McCarthy, Commandant of Fort Baker; and dignitaries from the press and the offices of Bethlehem and Roebling.

Most of those who gathered at midspan that crisp blue November afternoon were gritty, hardhatted ironworkers who had earned with sweat the right to a ringside seat. They watched as, at 2:00 PM, Joseph Strauss himself manned the controls of a stiffleg derrick. Strauss swung the lines over to a flatcar, where hooker-ons secured them to the designated chord. He shifted the controls, and the 100-foot chord rustled from the pile. Guyed by tag line men, it climbed high over the site, swaying gently as it moved toward the last gap in the steel. Then Strauss eased it down slowly until finally it was close enough for the connectors to grab it, position it, wave off the controls, make a fit, and slam some drift pins home.

The Golden Gate Bridge—the longest, most challenging suspension span in history—was finally in place. A Bethlehem connector captioned the moment for posterity. "That's the last piece, Bill," he said laconically at 4:25 PM.

Through the first forty-six months of the project, the workers on the Golden Gate seemed somehow charmed. In spite of the complicated interaction of people, machinery, weather and heights, they sustained remarkably few injuries. Men had fallen into the net and broken bones. They had been hit by drift pins or tools dropped from the higher reaches. They had been overcome by fumes in the towers and impaled on equipment in the foundations. They had been knocked senseless, made dizzy, cut, bruised, scratched, and burned. But the accidents were isolated, seldom serious, and never fatal.

That all changed about a month before the floor steel was pinned together. It became obvious, beginning on October 17, 1936, that fate had turned malevolent and no precaution, no matter how advanced, could completely protect the project. That day a bridgeman named Albert Zampa was saving a few steps on the Marin backspan by jumping from stringer to stringer. He missed his footing on a crossover beam and fell into the net at the spot it fanned against the craggy hillside. The net bellied onto the rocks, and Zampa absorbed the full shock of his fall. It broke his back. He was in critical condition for weeks and out of work for the next eight months.

Four days after Zampa's fall, a junction pin at the top of a stiffleg derrick pulled loose near midspan, dismantling the entire crane with an explosive bang that started the men scurrying for cover.

I was the signal man for the crane at the time. I didn't see anything at first, but I sure as hell heard it. That falling derrick sounded like Big Bertha going off. I looked up just in time to watch the boom fall out over the ocean side cable and into the Bay.

Then the mast came down, and one support leg fell down across the girder we were picking up. A hooker-on named Kermit Moore was climbing down off the steel—half his body was up on it, and his legs were dangling over the side. That support beam came slamming down right on top of him, hitting him across the head and shoulders and crushing him to death. The tag line man was knocked off into the net, but he survived.

You never know why those things happen, but on bridges they just do. It really shook us up. And as soon as things calmed down a little, the company sent us home for the day, like they always do when somebody dies. (Joe Walton)

Kermit Moore was the first fatality of the Golden Gate, and his death gave the project a more sober sense of destiny. For the remaining six months of work the bridgemen formed a "Halfway to Hell" club. Only men who had fallen into the net were eligible; by January 8 the roster numbered eleven.

On January 5 an accident occurred on the steel arch over Fort Point. The arch had just been connected and riveted, and the contractor, J. H. Pomeroy, was ready to dismantle the falsework scaffolding which had served as a construction platform.

To save money, Pomeroy commissioned low-paid pilebutts to remove the wooden X-bracing from the base of the support posts. They did this while steelworkers were taking bolts from the higher rungs of the scaffold. The pilebutts finished their task quickly, too quickly as it turned out. Without the X-bracing, the falsework could not resist a snarling wind that eddied in under the deck.

Up on top of the scaffold, three Pomeroy steelworkers felt it pitch and sway. "Let's get the hell out of here," one of them yelled, but there was no place to run. The whole structure collapsed, dropping all three men 150 feet into the courtyard of Fort Scott. Two of them, Pinky Brinkley and Bob Krieger, were only shaken up. The third, Jay Hollcraft, fractured his spine.

The funny thing is, we were working for Pomeroy only because we quit Bethlehem. I was signalman for raising steel near the San Francisco pier, and one day we was going slow because, for some reason, the crane cables had an uneven slack. This new foreman we had thought it was too slow. He came over, grabbed the piano boxes from my hand and told me to get the hell off the job. He started pushing buttons fast and furious, only he pushed what they called a whip line by mistake. That's the one with the headache ball attached, and when he did that, the ball went flying down, clear through to the bottom of the barge.

He could've killed somebody. The bosses down below saw the whole thing, and they asked me to stay on the job. But by this time I was really hot—I quit right then, and Bob and Pinky quit with me. That was about noon. We walked down the Bridge to Pomeroy's headquarters and told them we was three-quarters of a raising gang looking for work. They started us right away putting up the steel for that arch over Fort Point.

When we fell it was big news. I was in the hospital, so nobody got a picture of me, but Bob and Pinky had photos of themselves spread all over the front pages. Only problem was, Pinky had run out on his wife some years before. He never paid alimony, and she lost track while he was following steel all over the world. She lived in San Francisco, and as soon as she saw that picture, she sicked the sheriff on Pinky. I don't remember whether he had to go to jail or not, but I do know that at his first free minute he lit out again.

That created a problem for me. I wanted compensation while my back mended. I contended that the company was negligent for having those pilebutts take out the X-bracing while we were still on top. Bob and Pinky were my only two witnesses, but when I got out of the hospital I couldn't find either one. Pomeroy had sent Bob to a project up in Alaska, and Pinky was God knows where. (Jay Hollcraft)

Pacific Bridge returned to the construction site on January 19 for its second contract, the paving of the roadway, traditionally the least eventful phase of bridge building. Steelworkers laid rails for the cement carts. Carpenters built forms wedged between the stringers. Rebar crews put down reinforcing steel, and welders spot-sealed them, trotting so fast along the panels that the inspectors had difficulty keeping pace. From there it was the standard routine of mixing, pouring, vibrating, leveling, and curing the cement—on a much magnified scale a job no more complex than the construction of a suburban driveway.

It should have been an effortless last lap, but twenty-eight days after Pacific Bridge initiated the contract, the project was shattered by misfortune and grief.

To remove the wooden forms, men had to work underneath the deck. For that Pacific Bridge designed and installed a traveling platform that dangled from the underside of the floor girders. It moved by means of a hand-turned winch, and it was connected to the girders by four clamp-like fixtures that resembled ice tongs.

Men were supposed to pull out the cement-flecked pieces of wood and send them up a bucket brigade to the roadway. When they finished stripping the wood from one panel, they were to release the winch and crank the stripping stage to its next stop. But many doubted the soundness of the scheme. On the first day of form stripping, the crew climbed to the staging, but with serious misgivings.

When we first saw that scaffold we noticed things that didn't look quite right. There were a bunch of bolts to put through the pincers of these tong arrangements so they couldn't open up due to any transverse friction. The bolts were a necessary requirement, but the ones they brought out were too goddamned short. I told somebody up there—I think it was Shorty Bass—that I didn't think this platform was a damn bit safe. (Wayne DeJanvier)

After investigating the platform, District safety engineer Al Maillioux also felt uneasy. Maillioux had no authority to close down an operation, only to make recommendations. Accordingly, he contacted Pacific Bridge and asked officials to meet him at the site at 9:00 AM on February 17, the morning the platform was to be used for the first time.

They showed up on schedule, and as they were walking toward midspan, they were stunned to a halt by earsplitting, staccato sounds.

There are so many stories about the staging collapse, most of them hearsay. I know what happened. I was there. I had just been on the scaffold looking for a guy. I couldn't find him, and my buddy Slim Lambert came over and we started talking, I don't remember about what, probably what we were going to do on Saturday night.

I decided to go up to the deck and look for the guy I was supposed to find. I was still talking to Slim as I climbed the ladder. I got on top, then lay on my belly on the roadway to continue this conversation. Slim said something, but I didn't hear it. So I leaned closer, cocked my head to one side, cupped my hand to my ear, and asked, "What'd you say?" I didn't get an answer. When I looked down again, Slim wasn't there. Nothing was, no men, no tools, no net, no nothing. You couldn't hear much, what with all the noise on the bridge, the rivet guns, the compressors, the wind and the like. All I heard was somebody yelling, "Run Andy, Run!"

Then I saw. The platform had collapsed, spilled into the net, and broke it. The net had fallen the 200 feet into the water and got caught in the outgoing tide, which was pulling it so taut that Chris (Andy) Anderson, who was in the net at the time, could run up it like a slide. But the net was coming apart at the outriggers too—Bing! Bing! Bing! at the same time Andy was running like hell. Then, all of a sudden, Andy stopped running and looked up at us. Right then, the net broke completely, wrapped him up, and took him into the hole. We found him three or four days later still wrapped in a section of net. (Pete Williamson)

The noise that Maillioux and the Pacific Bridge people had heard was the net popping loose from the outriggers. The stage tore out several sections, and the force of the tide kept ripping it ragged at its weakest point, the manila hemp junction between the mesh and the steel rods. Over 2,100 feet of net, the distance from midspan to the south tower, was torn free. And when witnesses regained their composure they found that twelve men had gone into the hole.

Slim Lambert somehow survived the fall—he was one of two who did. A 51-year-old carpenter named Oscar Osberg was pulled alive from the water with a fractured hip, a broken leg, and massive internal injuries. Lambert suffered no serious effects. When he was rescued by a fishing boat, he was swimming strongly with the tide, clutching the body of 24-year-old Fred Dummatzen. Dummatzen, alive after he hit the water, was killed by plummetting debris.

Nine men died with him: Chris Anderson, and another Anderson, a 41-year-old laborer named C. A.; Shorty Bass, 37, laborer; Orril Desper, 35, laborer; Terrance Hallinan, 29, laborer; Eldridge Hillen, 33, laborer; Charles Lindros, 29, carpenter; Jack Norman, 28, laborer; and Louis Russell, 22, laborer. "Ten of my friends," Lambert said days later from his hospital bed. "I saw them die all around me, and there wasn't a damn thing I could do about it."

Up on the deck, shocked workers were busy hauling up two men who had grabbed onto girders just as the platform started to tilt. Their yells brought rescue lines, and both were reeled in to safety. Newspaper accounts focused exclusively on one of them, an affable pipesmoking Irishman named Tom Casey. The other was Wayne DeJanvier.

I was up there all right. If you want proof, go look at the girder I was hanging from—I'll bet my fingerprints are still imbedded in it.

I was at the front of the platform when the back end started to tilt. That teetered me right up to the floor beam. I reached to grab the flange, and I just hung on. I was hanging there by my hands for a while, yelling like crazy. As soon as I could think, I swung around and grabbed the flange with my legs. Then I held on, one hand at a time, while I pulled off my gloves with my teeth and spit them into the hole. All this time I was still hollering. Finally, somebody—I think it was Ken Wellman—heard me. He and some of the other boys threw a hawser down. They swung it back and forth like a pendulum until I caught it.

That kind of thing really shakes you up. I couldn't stand to be there so I left right away. I knew a guy who owned the corner bar near my apartment. I went straight there and killed a couple of bottles. Then my roommate came in, and we bought some bottles to take home.

In the process of killing them we got pretty well inebriated, and finally we went to sleep. Some time in the middle of the night, there was a loud rapping at the door. I opened it, and four or five reporters came barging in. They started flashing pictures and asking a bunch of stupid questions.

Here we were, half asleep, half drunk, completely depressed, and standing there in our longjohns. I took one look at the situation and told my partner, "Let's get rid of these assholes." He nodded, so I grabbed a camera and threw it out in the hallway. Then we both boosted these guys out of the room. That's probably why I never got my name in the papers. (Wayne DeJanvier)

Within 24 hours a coroner's jury was inquiring into the death of the first recovered victim, Fred Dummazten. Resident engineer Russell Cone testified that the safety bolts were in place when he had inspected the assembled scaffold, but several workers claimed that the bolts were missing. A day later, on February 19, the jury reached its verdict: "We . . . find . . . death due to the spreading of hooks . . . evidently through the failure to use certain safety equipment, namely a bolt."

For weeks after the accident, recrimination went whipping through the Bridge site. Neither the District nor the Industrial Accident Commission could fix a cause, and nobody volunteered to take the blame. Many of the workers thought privately it was a case of "highball and haywire," rushing through a job without proper precaution. After an internal study conducted by Clifford Paine, Joseph Strauss issued the District's first public statement. He bluntly accused Pacific Bridge of negligence. "The cast aluminum side plates were seriously in error," he said in part. "Failure was inevitable."

Pacific Bridge charged in acrid reply that the District's engineers were shopping for scapegoats. "It would seem," intoned company president Philip Hart, "that Strauss and Paine are attempting to evade responsibility and divert blame for their failure to exercise . . . supervision over work in their charge."

The Industrial Accident Commission subsequently said the accident was owing to undersized safety bolts, but it held nobody liable.

Gradually nerve endings uncoil and tensions subside. Work proceeded slowly for the next several weeks. On March 3, the replacement of the net began; little by little the men resumed their former pace. As the final squares of pavement were set down that spring they began to express gratitude and relief that they had beaten the one-man-per-million-dollars ratio by twenty-four lives. And they began to feel proud. The most beautiful bridge in the world was nearly complete, a monument to progress that all of them had helped to construct. ■

Roadway construction began at the towers and moved out in both directions at once. Travelling cranes handled the steel pieces for the raising gangs, and the gangs connected them to the suspender ropes. Before each new panel was started, the safety net was extended, so it was always well in front of the workers.

Left
One morning the wind was so strong you could see the net billowing at something like an eight-foot pitch. The boss said a lot of tools had been blown down into the net, and he told me to go down there so he could lower a line for hauling all the stuff back up. Well, I retrieved everything, but it was so nervewracking that when I returned to the deck I was soaking wet from my own sweat. (George Susuoeff)

Right
The suspension principle was in force when the main cables began to bear the weight of the roadway. That happened as soon as the steelworkers inserted the suspender ropes into the channeled flanges of the floor system steel and locked their weighted bucket-bottoms in position with gusset plates.

The main longitudinal members of the roadway were called "chords." Top and bottom chords, separated by vertical and diagonal beams, formed the "trusswork," the outer faces of the floor system.

The net extended well beyond the width of the roadway, which was fortunate for the steel-workers, who often rode on a trusswork piece while the travelling crane swung it into position.

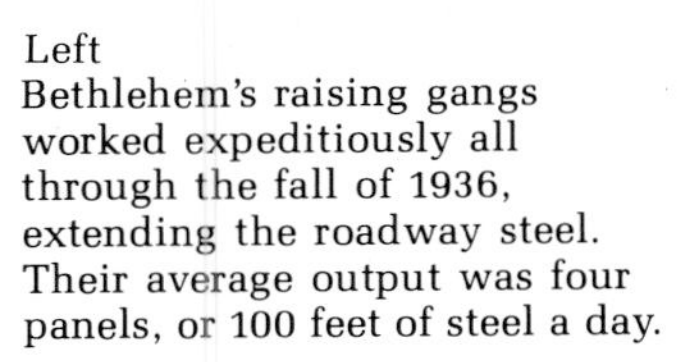

Left
Bethlehem's raising gangs worked expeditiously all through the fall of 1936, extending the roadway steel. Their average output was four panels, or 100 feet of steel a day.

Right
A lot of the old bridgemen have hearing problems, and the reason is the deafening noise we worked with every day. You couldn't conduct a normal conversation out there, even with somebody close by. You'd be drowned out by the sounds of rivet guns, derrick engines, compressors, steel slapping against steel, and the wind.
(Pete Williamson)

The chords were actually hollow box girders 2½ feet in circumference. To "buck" the chord rivets, the smaller ironworkers had to crawl inside them with a "wildcat jam"—often toiling entire shifts at a time in those dark and narrow enclosures.

Left
As the panels progressed, the floor began to take shape. The longitudinal strips set inside the chords were called "stringers." The stringers were part of the floor system, but they also served as rails for both the flatcars carrying steel to the cranes, and for the cranes themselves.

Right
Like the perilous task of riveting sideplates on the chords, much of roadway work was performed by men perched awkwardly on narrow steel beams. For that reason, they were happy to have the net. "It allowed us to work with greater abandon," recalls one ironworker.

On November 18, 1936, the last 90-foot floor beam was lowered into place and the main span was closed. Bridge District directors and engineers gathered to witness the milestone event.

The painters had to treat all areas of exposed steel, particularly the hidden corners and angles where moisture was likely to settle. Sometimes the job required the skills of a contortionist.

In raising the roadway, the four derrick crews had to coordinate their efforts so the cables were always bearing the same proportion of weight. As a result, the normal bow of the bridge appeared exaggerated in the beginning and middle stages of roadway construction.

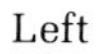

Left
There was always four men in a rivet gang, counting the heater. He'd be cooking the rivets, then when it was time, he'd toss one to a catcher, who'd grab it out of the air with his bucket. Then, the catcher'd take it out with his pickup tongs and back it into the hole, so's the buckerup could jam against it with his collybar and the riverter could form a head with his rivet gun. (George Susoeff)

Right
The floor system proved the most trouble-free phase of bridge construction. It was begun on September 11, 1936, and with Bethlehem's experienced ironworkers putting up as many as 640 tons of steel a day, it was completely connected and riveted 94 days later, on December 14.

As early as 1934, approach roads from San Francisco and Marin began snaking their way toward the channel. By the first of the year, 1936, the steelwork on either side of the channel was in its final lap—the distance between the hillside and the pylons. The Marin approach was finished first, in August of 1936. The San Francisco side, pictured here, was complicated by 200-foot high support towers, and by the 319 foot arch built over Fort Point. It was not completed until late in the fall of 1936.

Left
The span was near completion by the beginning of 1937. All that remained were the paving of the roadway, the dismantling of the catwalks, and incidentals, like the continuous painting of the suspender ropes.

Right
Working on the suspender ropes, the painters dangled from bosun's chairs for hours at a time—so long that their legs would often fall asleep. For relief, they would periodically disengage from the hempen chairs and hang outstretched from their safety lines.

Left
When we spot-welded the reinforcing iron for the highway concrete, we moved across it like turkeys picking corn—hunkering down, wiping the point with a fast-burning rod, then rushing to the next position. We worked so fast the inspectors couldn't keep up.
(Whitey Pennala)

Right
Roadway concrete was trucked along the construction wharf to a heavy-duty hoist at the base of the south tower. From there it was raised to a roadway-level hopper, which dropped it into industrial dump cars. Once loaded, the string of dump cars chugged to the work sites behind a small gasoline locomotive called a "donkey engine."

Left
Prior to pouring, Pacific Bridge workers installed wedges perpendicularly between the stringers. On them they laid sheets of plywood, which would be the solid fundation for both the reinforcing iron and the concrete.

Right
The roadway was poured in three 20-foot-wide ribbons, the outer lanes finished first. Concrete was set down in 50-foot lengths, and between each length workers installed steel-and-copper expansion joints. These would alleviate any strain on the pavement from the anticipated contraction and expansion of the floor system steel.

By April 15, 1937, the paving—the last of the structural chores—was finished. In spite of unforeseen delays, the Bridge had taken only 4½ months more to build than had been originally planned. Even more amazing was the absence of cost overruns. In 1930, Joseph Strauss had estimated construction costs at $27,165,000. When finally totalled, they were actually less—$27,125,000 (not including easement, financing, engineering, and administrative fees which brought the final figure to $35,000,000). The Bridge had fulfilled its promises, and now it was just one massive cleanup operation away from being open for traffic.

I was one of the first people ever to drive across the Bridge. It happened about two weeks before opening day, when this guy "Tobacco George" Buttner and I were helping dismantle equipment around the north pier. George got banged up pretty bad by some falling cables—he was unconscious and had some broken bones—and I had to bring him to shore. I found a gurney, strapped him in, and managed to get the gurney inside my Dodge convertible, which was parked nearby. Then I called the field hospital over at Fort Point. They said, "Bring him over." I asked, "How?" They said, "Drive!" So I did, weaving around all kinds of junk and equipment that was still laying on the deck. They wouldn't let me drive back, though. I had to take the ferry. (Gerry Conser)

OPENING DAY: THE DREAM IS REAL

Here is your bridge, Mr. O'Shaugnessy.
—Joseph Strauss

Ceremonies of Completion

Pacific Bridge finished paving the deck on April 19, 1937. Nine days later the first ceremony of completion was held near midspan. A Sonora man named Charles Segerstrom had cast a rivet from historic High Sierra gold and donated it to the District. Now, under matchless blue skies the morning of April 28, it was going to be driven into the span as the symbolic last rivet.

Shortly after 9:00 AM two companies of men from Fort Scott's 6th Coast Artillery joined ranks behind their company band at the toll plaza. They lined up triumphantly the width of the Bridge and set out marching toward centerspan. As they reached the first pylon they were met by a horrified Pacific Bridge official, obviously alarmed by so many close-order feet tromping over his curing deck. "Have your men break step," he screamed at the band leader. "Do you want to ruin the Bridge?" The entire command fell into disarray and, deflated, shuffled toward the ceremony.

At midspan, with hundreds of people in attendance, the agenda began with the usual string of political speeches. Joseph Strauss likened the moment to the driving of the Gold Spike at Promontory Point sixty-eight years earlier. Merrill Brown, chairman of the Golden Gate Bridge Fiesta Committee spoke passionately of the beauty and utility of the span. Mayor Rossi invited all the world to come to San Francisco a month later to help celebrate the official opening of the world's longest and most beautiful single-span suspension bridge. District President Filmer and Fort Scott Commandant Harold Cloke echoed their sentiments.

Finally, John F. Shelley, president of the San Francisco Labor Council, handed Segerstrom's gold rivet to ironworker Edward Stanley, the man who had driven the first rivet on the Golden Gate, the tenth member of the "Halfway to Hell" club. Teamed with buckerup Ed Murphy and accompanied by cheers, Stanley jammed in with his gun. Unfortunately, his touch was hardened from too many years of driving nothing but steel.

There was no forge out there. They figured that since it was gold it was soft enough to drive cold. So they just hooked up to a gun and a dollybar, and Murphy, he bucked it up. When Iron Horse Stanley went to drive it, he had trouble forming a nice head. He fussed and fumed and sweated. He'd drive for a while, then stand back and look at it, then drive some more. He stayed on it so long that Murphy raised a blister on his palm the size of a half dollar.

When they still couldn't get a good head on it, they took a burning torch and tried to heat it. They heated it for a long time, too long I guess. When Stanley drove it again, the head fell apart. He pulled his gun back, and all the pieces dropped to the ground. Everybody was out there that day, brains from the Bridge and a lot of civilians and officials, and Iron Horse Stanley stayed on it too long. (Louis Hack)

Stanely's rivet hole was later filled with steel (a ceremonial gold rivet was eventually inserted several panels away), making the Golden Gate Bridge structurally complete. All that remained were finishing details, random painting and clean-up, some ornamental iron work on the tower portals, the installation of the toll plaza equipment—and completion of plans for the celebration fiesta to be held the week of May 27 to June 2.

District officials expected the largest crowd ever to congregate in and around the city of San Francisco. Delegations had been invited from as far north as Alaska and as far south as Guatemala, from all eleven Western states, from each of California's fifty-eight counties, from every conceivable military and service organization.

For months prior to the celebration, Boy and Girl Scouts planted and nurtured blue lupin and golden poppies in the bare soil around the bridgeheads. A composer was commissioned to write a theme song. A fiesta queen was chosen from scores of candidates. A specially appointed committee arranged for nightly illumination of the Bridge with strings of colored lights running the entire length and height of the span. "We will present it as a luminous jewel in the sky," claimed committee chairman Tirey Ford.

There was more: whole streets, notably Polk and Market, were costumed in Gold Rush decor; officials invited contestants for a San Francisco-to-Marin Golden Gate swim; a lighted parade involving 150 floats and 15,000 marchers was planned for the night of May 29; a week long pageant entitled "Span of Gold" and requiring the participation of 3,000 players was created and choreographed for performance in Crissy Field's Redwood Grove theater. Clearly, San Francisco was ready to whoop: "Never since the war have we seen the old town so in the mood to cut loose and celebrate just for the joy of celebrating," said a May 26 editorial in the *Chronicle*.

The District planned for such an elaborate array of events, invited and anticipated such a sea of celebrants, that an auto trailer camp had to be established at the corner of Ocean and Phelan Avenues in the Ingleside District. And very early in the planning it became obvious that a single opening day would not be enough.

I was on the committee that was planning the celebration for the opening of the Bridge. I did a little research and brought to the committee one day some tearsheets from the New York Times which had a picture of the Brooklyn Bridge in 1883, on what they call "Pedestrian Day." It showed a mob scene of people. I suggested that was the thing to do for the Golden Gate—have two celebration days, one for pedestrians, and one for automobiles. After about two months of negotiations that idea was approved by the directors. And Pedestrian Day turned out to be the biggest thing about the opening of the Bridge. (Ted Huggins)

By foot, by automobile, by motor caravan, by public transportation, even by rickshaw they came to the bridgeheads for Pedestrian Day on May 27. There was a heavy schedule of collateral activities, including a hard rock drilling contest and a fireworks display at Crissy Field and an industrial and manufacturers exposition at Dreamland, but the main event was the first pedestrian crossing of the completed span.

The day dawned chill and foggy, but excitement burned almost visibly through the morning mist. By first light an estimated 18,000 people were already at the site. Inquiring reporters located a Boy Scout named Walter Kronenberg at the head of the pack and identified him as the first in line. He had been there, he said, since 7:00 PM the night before.

Buoyant mobs from both sides of the Gate awaited the opening signal. At precisely 6:00 AM the foghorns at the center of the Bridge bleated and barriers fell. District officials had hoped for orderly crossings, people parading down one sidewalk from Marin and up the other from San Francisco. But things turned quickly chaotic amidst excessive numbers and unbridled exuberance. From the moment the fences were down some pedestrians raced across, some gamboled, and others gave in to the welling instinct to stand and pay homage.

It was a day for firsts. Donald Bryant, a runner from San Francisco Junior College, was the first to sprint across the Golden Gate. Some young boys without the nickel pedestrian toll were the first to negotiate a crossing along the hempen route of the safety net still slung underneath the roadway. Two sisters were the first to roller skate across. There were also the first baby in a stroller, the first priest, the first father and daughter to carry 25 pounds of Schuylkill County (Pennsylvania) coal across the span, the first person to walk the distance with her tongue sticking out, the first dog (a Scottie) and the first person to cross the bridge on stilts (a woman).

Meanwhile scores of people congregated on the roadway. Some listened excitedly to the wind strumming eerie melodies on the suspender ropes. Some stood frozen in reverie. The more adventurous frolicked, sometimes perilously, along the railings, drinking in the scenery. Others just threw blankets on the deck and sat down for picnics.

An average of 200 people per minute entered the turnstiles of the Golden Gate Bridge that breezy day in late May. With approximately 200,000 pedestrians crossing in all, it was a minor miracle that no serious misfortune occurred. The two ambulances and sixty-five policemen on call that day attended to the usual run of blisters, an occasional broken bone, and the inevitable lost children. But overflowing with good will and pride, Pedestrian Day was a happy and, on the whole, trouble-free occasion.

The next day was the second opening, this one for automobile traffic. The event was as significant as Pedestrian Day, but it was more staid. This was the official, not the popular opening, and it was as different from the previous day as a motorcade for heads of state is from a Fourth of July picnic.

Very early in the morning the official party headed by Mayor Rossi and Chief Engineer Strauss and including engineers, District officials, politicians, and representatives from Canada and Mexico assembled at Fort Point. From there they were ferried to Sausalito and transported to the Marin bridgehead by automobile.

At exactly 9:30 AM they began to make their way through a series of ceremonial barriers. The first they encountered at the near pylon, a sixteen-foot-long redwood log almost a yard in diameter. At a signal, the log bucking champions of the Pacific North Coast competed to saw through the barrier. In two minutes and 47 seconds Paul Searles of Longview, Washington opened the way for Rossi's party and received a $500 prize for his efforts.

At the second pylon the path was obstructed by three chains of precious metal. An acetylene torch was handed to the "Father of the Bridge," Franklin Pierce Doyle, and he cut through the copper chain in minutes. He passed the torch to Mayor Rossi for cutting through the silver; Rossi passed it to District President Filmer for cutting through the gold.

The Marin end of the Bridge was now open. Led by the official cars, hundreds of vehicles streamed slowly toward a final barrier, a picket line of fiesta queens guarding the entrance to the toll plaza. They would not give way until Chief Engineer Strauss formally presented the span to President William Filmer. As he handed over the Bridge to Filmer, Joseph Strauss quivered noticeably. For twenty-one years he had been synonymous with the Golden Gate Bridge. More than anybody he was responsible for this magnificent structure. More than anybody he should have been allowed to speak in boastful tones.

But the solemnity of the day muted even this proud and sometimes pompous man. Barely audible through the unrestrained cheering of spectators, his voice cracked with emotion. "This bridge needs neither praise nor eulogy nor encomium," he said finally. "It speaks for itself."

With that he receded into the crowd, his life's work opening for the ages behind him.

There was one more official opening event. Precisely at noon, President Franklin Delano Roosevelt, in the White House 3,000 miles away, pressed a telegraph key announcing that the Bridge was finally in public use. At that moment the sky erupted. Five hundred aircraft from the Navy's carriers at sea flew overhead while every foghorn, churchbell, police and fire siren, ship's whistle, firework, automobile horn, and it seemed every noisemaking device in the entire San Francisco Bay Area shrieked in cacophonous accord. The Golden Gate Bridge was open. ■

On April 28, 1937, two weeks after the paving was completed, Joseph Strauss led a group of dignitaries out to center span, where buckerup Edward Murphy and riveter Ed Stanley would drive a ceremonial gold rivet. Unfortunately, there was no forge on the Bridge at the time, and Stanley tried to drive the rivet cold. He drove so long and hard that he eventually dissolved the rust into flakes, embarrassing himself and aborting the festivities. The hole was later filled with a conventional rivet, and a gold one was unceremoniously driven into another part of the span.

Right
Following the example set by the Brooklyn Bridge 54 years earlier, District officials preceded the opening day for autos with "Pedestrian Day." The gates would open at 6 AM on May 27, and by dawn, thousands of foot passengers had assembled at the bridgeheads.

Left
An average of 200 people a minute entered the span in the 12 hours it was open for Pedestrian Day. By the closing time of 6 PM, more than 200,000 revelling citizens had crossed the Bridge.

Right
Some stood and stared, others walked directly across, but all were struck by the enormity of the day. "It was a big thing, that bridge, the biggest thing in the world," said one of the directors, "and those people who gathered to claim it were proud that it was big. . . ."

May 28 was the official opening day for autos. Early in the morning, Navy planes flew in formation overhead while curious bystanders stood vigil awaiting the first wave of traffic —a motorcade of dignitaries that soon would be crossing from Marin.

At the Marin end of the span, the ceremonies were underway. Near the second pylon, the motorcade was halted while the "Father of the Bridge," Franklin P. Doyle, severed a copper chain, the first of three precious-metal barriers to be cut by Bridge officials.

The final barrier was a line of fiesta queens guarding the toll plaza entrance. They would not break ranks until Joseph Strauss had strolled to the microphone and formally presented the span to District president William P. Filmer.

Led by a synchronized escort of police, the motorcade proceeded slowly and proudly toward San Francisco.

Causing traffic jams on Highway 101 to the north and in the city of San Francisco to the south, autos lined up three abreast at the Waldo, Presidio and Marina approaches. At precisely 12 o'clock noon, President Franklin D. Roosevelt punched a telegraph key in the White House. In an instant, a small electrical impulse shot 3,000 miles signalling that the Bridge was open. The news was relayed quickly and backdropped by sirens, whistles, horns, bells and cheering. The cars came streaming across.

EPILOGUE: THE BRIDGE

When you build a bridge, you build something for all time. —Joseph Strauss

Almost two decades had passed since that day in 1919 when the bridge first was conceived, when the tiny engineer from Chicago, Joseph Strauss, had stood on the windy bluff overlooking Fort Winfield Scott, reconnoitering the site of future triumph, looking through a stereopticon, as it were, at a scene in sepia tones he fully planned on dissolving into technicolor.

It was a year later in 1920, in moments stolen from a busy schedule, that poet Joseph Strauss penned something that had been stirring inside: a vision of California he no doubt had recognized the year before when gazing across the unspanned gorge.

"California, land of wonders;/Land where the Pacific thunders/...Where white Shasta seeks the skies,/Home of the mighty Tamalpais/...Land where summer zephyrs blow,/'Neath the winter peaks of snow;/...There, where the Pacific thunders,/California, land of wonders."

Spanning the Golden Gate was never just a project nor a job for Joseph Strauss. It was a life's purpose. The bridge he imagined in 1919 was a man-made wonder blending seamlessly into a setting of natural wonder. Two decades later in 1937, the Bridge was built, the traffic was flowing, and the vision was real.

Now headed toward its second fifty years, Strauss's mark is still indelible: his bridge is so much a part of the wild scenery of the Golden Gate no one can imagine the channel without it. More than anything he ever put on paper, the bridge is Strauss's true poetry, his immortal imprint.

Joseph Strauss retired officially as chief engineer of the Golden Gate Bridge and Highway District in early summer of 1937. Sixty-seven years old, he would undertake no more projects. He returned to Los Angeles where he lived with his second wife. Within a year, after being bedridden for two months, he was dead of a coronary thrombosis.

"The bridge killed him," said his elder son Ralph, a retired Army Colonel, in 1986. "You'd have to say that, the strain of it all. I'm only sorry my father never lived to see the acclaim the bridge has had...it was so difficult when he was trying to promote that undertaking. Everyone was denigrating him."

Strauss's vindication began, of course, with the triumphs of opening day. "The Golden Gate Bridge," he said, "the bridge which could not and should not be built, which the War Department would not permit, which the rocky foundation of the pier base would not support, which would have no traffic to justify it, which would ruin the beauty of the Golden Gate, which could not be completed within my cost estimate of $27,165,000, stands before you in all its majestic splendor, in complete refutation of every attack made upon it."

Strauss would continue to be vindicated. The Bridge has endured a half-century test of time and weather with no sign of faltering. "No span of steel will tolerate...neglect," he said. "But if serviced by the generations who use it and spared manmade hazards, such as war, it should have life without end."

From the beginning, maintenance has been meticulous. Bridge painting, for instance, grew from guesswork to science on the Golden Gate. Russell Cone had said in 1940, "The painting of steel bridges has been neglected for a long time. People are just now waking up to the fact that it's not only a wise thing to do but it's essential. It's what is known as protecting the investment."

At the onset, bridge painters used the traditional red lead system. They worked every day that weather permitted, routinely losing 90 days a year to inclement conditions. Sitting in bosun's chairs that hung from wooden scaffolds, they chiseled away blistered remains of peeling paint. Consuming 500 gallons a month, they brushed on thick coats of red lead, swabbing over the rust, the millscale, and the dirt. But over time the red lead system became a losing battle. Decades passed, and corrosion was eating through the bridge steel faster than the painters could arrest it.

Starting in 1968, they put away their brushes and turned to atomizing and airless sprayguns. They also retired the red lead paint in favor of an inorganic zinc substance first developed by the U.S. Navy for use on submarines. They tried it initially on the south tower in 1968, sandblasting down to bare metal and spraying on the inorganic zinc. The technique proved as durable as galvanizing. Zinc substance is "sacrificial"; when sprayed on the bridge, zinc molecules blend into the surface, "sacrificing" themselves to protect the steel molecules.

Bridge painters tested the zinc on the south tower, then they converted the whole system. The new process, sandblasting and zinc recoating of the entire span, was expected to take the 39-man painting crew 25 years. At the time of the golden anniversary in 1987, the conversion was estimated at more than 80 percent complete.

Beyond the matter of regular maintenance came the big tests. On February 9, 1938, for instance, a storm blew so fiercely through the Gate it was impossible for a pedestrian to stand erect on the sidewalk. The chief engineer—it was Russell Cone then—noted the bridge had leaned off center as much as eight to ten feet. Three years later, on February 11, 1941, a 60 mile-per-hour wind roared in from the southwest, causing the towers and roadway to bend from their axes by nearly five feet.

The bridge remained well within its tolerances, so no one was alarmed. It was truly the "rubber" span that Strauss had joked about in the War Department hearing of 1924. But in 1941, bridge engineers became aware of a new danger inherent to suspension spans, something they called, "harmonic oscillation."

A storm had swept through Tacoma, Washington, in 1940, proving that certain winds could flutter a roadway into self-perpetuating waves so vigorous they could ultimately rip the floor system free of the towers and topple the entire structure. The evidence was undeniable: the newly opened Tacoma Narrows Bridge lay as a pile of rubble at the bottom of the strait.

The Tacoma Narrows was the third largest suspension bridge in history when completed in July of 1940—2800 feet between its 425-foot high towers. It surpassed the Golden Gate in sleekness of profile, having the world's smallest ratio of width to length. It was this delicate form plus a solid girder system preventing wind diffusion that caused its failure.

With lace girder chords and open trusses, the Golden Gate was certainly able to deflect the force of the winds. But the Tacoma disaster prompted new interest in the ratio of width to length. The Tacoma Narrows Bridge was built at 72 to 1, the Golden Gate at 47 to 1. "For this reason," declared Russell Cone at the time, "the Golden Gate span must be regarded with suspicion."

No more was mentioned until December 1, 1951, the day a 69 mile-per-hour gale howled through the gorge, quivering anemometer dials and shaking the towers down to the piers. The storm was so severe the roadway started to ripple. One side was pitching 11 feet higher than the other, when bridge officials closed the span for the first time in its 14-year history.

Nearly three hours later, the winds had subsided, the roadway waves began to recede. Casual inspection the next day revealed little damage—some suspender ropes had pulled their socket plates out of shape; some light standards were bent back virtually to meet the roadway. For a more detailed report, the District asked Clifford Paine to come west.

Paine found only one suspect area, the lateral connections between the floor system and towers, which were dangerously contorted. He prescribed the necessary repairs, and he made a recommendation. Having headed the committee which studied the Tacoma Narrows failure, Paine now knew more about the effects of wind-induced oscillation than anyone. To insure against a Tacoma-type mishap, he said, the bridge would need a series of stiffening girders criss-crossed between the chords underneath the full length of the roadway.

The idea was accepted, and in the mid-1950's, the $3.5 million job was awarded to Judson-Pacific Murphy of South San Francisco. It was one of only three major modifications the bridge has known since it opened. Another was made in the early 1970's when all the suspender ropes were replaced—after corrosion had been discovered near the gusset plates at the junction of some ropes and the floor system chords.

The American Bridge Company handled the problem. Crews brought in huge jacks, which worked much like vertical vises. The lower jaw was set under the floor chord, the upper on top of the main cable. Downward pressure from above and upward pressure from below narrowed the distance between the roadway and cable, loosening the tension on the suspender ropes. The ropes then were pulled off and replaced with temporaries, while workers refashioned the chord sockets before installing new ropes. American Bridge needed nearly four years to complete the task.

The roadway was next. It was suffering wear from almost four decades of heavy use. From opening day to the early 1970's, more than 500 million cars had crossed the span, which projected to more than a billion by the time of the fiftieth anniversary. Traffic was much thicker than Strauss ever had imagined. It was creating more stress than the roadway could withstand indefinitely.

By the time of the fortieth anniversary in 1977, the deck was showing a pattern of cracks made ominous by conditions in the channel. Fog and salt air that blow in daily were sifting through the roadway cracks and depositing salt chloride into the grain of the concrete. Bridge engineers were concerned. Chloride content in the roadway concrete had reached three pounds per cubic yard. Residue from chloride and rust had increased the deck's weight and actually raised its level. Just as serious, the chloride had migrated through the cracks, to the reinforcing steel below.

Repair was impractical. The only solution was building a new roadway in place of the old. The district awarded a $52.5 million contract in November of 1982 to Dillingham-Tokola. The project was scheduled to take 400 days. Crews worked mainly at night. They closed down two lanes at a time and sectioned the roadway into pieces measuring 15 by 50 feet, about 800 of them. They cut through the seven-inch thick paving a section at a time, removing each piece and installing a new one in its place. The last structural section was put into place about 3 a.m. on August 15, 1985—401 days after work had begun. When the last step was completed—the two-inch layer of epoxy paving—the new roadway was stronger, more durable, and lighter by some 11,350 tons.

In all other respects, the bridge has stood solidly since opening day. American Bridge engineers cut sample wires from the strands inside the main cables in the 1970's, and they found no weakening whatever. The exposed girders are inspected regularly; they remain sturdy and free of rust, seldom in need of even minor repair. The channel around the piers has been surveyed, and the giant footings have stayed impervious to endless pounding from the tides.

Joseph Strauss achieved what he set out to achieve when he first envisioned the span—a bridge for his time and for all time. The Golden Gate Bridge in the decades since construction has aged with grace and strength. Strauss himself stands sentinel in the form of a green-bronze, life-size statue on the lawn near the roundhouse. He faces the city, his life's purpose rising behind him, in most respects still the technicolor imprint of that dream he had in 1919.

In most respects, Strauss no doubt would be surprised by some of what he would find today. Fifty years ago, for instance, he thought six ten-foot lanes of traffic adequate, possibly excessive. He would not be prepared to see all six lanes now choked with traffic: cumbersome RV's alongside sleek foreign convertibles, high-boy pickups chugging behind Volkswagen vans, Volvos, Mercedes, Fords, and Chevrolets. Traffic became so dense by the late 1970's that the curb lanes had to be widened by a foot during roadway reconstruction.

"Elbow room for trucks, buses, and recreational vehicles," says Daniel Mohn, project engineer for the District.

Strauss might also shudder at the chain link fence near the pedestrian walkway, or at the occasional graffiti spray painted on the railing, or at all the signs of warning and prohibition: "Injury or Defacement of this Bridge is a Misdemeanor," "Bicyclists Should Walk Bicycles on Bridge Sidewalks," "Skateboards, Roller Skates and Motor-Driven Cycles Prohibited on Bridge Sidewalks." He certainly would be aghast at the more than 800 suicides who have used the Bridge as their weapon.

"The Golden Gate Bridge is practically suicide-proof," he had said in 1936, a year before opening day. "The guard rails are five feet, six inches high and are so constructed that any persons on the pedestrian walkway could not get a handhold to climb over them..."

Joseph Strauss was wrong about that, of course, but right about so much else. Smog rises from the roadway, traffic congests the lanes. Politics has taken hold as usual. But Strauss's original conception remains alive—a bridge across the Golden Gate that stands with the dignity of age despite the scars inflicted by progress and time.

It is the bridge that reassures air travelers as they wend west to the Pacific, turn east toward the Atlantic, or make their final approach returning to San Francisco International Airport. The bridge that inspires sailors and mariners cruising out to sea or returning to home port. The bridge that is, among other things, the most obvious and cherished symbol of the City of San Francisco.

Joseph Strauss said in a radio address in 1930: "A great city with water barriers and no bridges is like a skyscraper with no elevators." He said, "Bridges are a monument to progress."

The bridge Strauss conceived that windy day in 1919 is that plus more. An enduring symbol of progress, of course. But also a monument to those who made it real, himself and the workers and engineers. It is their bridge and our bridge. A monument standing tall at its golden anniversary and beyond... ■